The Shrub Identification Book

THE
Shrub Identification
BOOK

The Visual Method for the
Practical Identification of Shrubs,
Including Woody Vines and Ground Covers

By

GEORGE W. D. SYMONDS

Photographs by A. W. Merwin

William Morrow & Company, New York

Copyright© 1963 by George W. D. Symonds

All rights reserved

Published simultaneously in the Dominion of
Canada by George J. McLeod Limited, Toronto

Printed in the United States of America

Library of Congress Catalog Card Number 63-7388

3   4   5   75

To:
ROBERT
GEORGE
DAVID
FRANCES

ACKNOWLEDGMENTS

The photographs in this book are all by A. W. Merwin of Wilton, Connecticut, and speak for him better than I can. I want to add, nevertheless, that in addition to his splendid photography, his sustained interest and patience under difficult conditions are greatly appreciated.

The assistance given to me by Mr. E. J. Alexander, of the New York Botanical Garden, both in the field and during many hours of consultation, was invaluable. I can only thank him for his generous and skillful guidance and say that I have seldom met anyone with such extensive knowledge hidden beneath such an unassuming and charming exterior.

I particularly wish to thank Senator Thomas C. Desmond of New York State for his kindness in giving me complete freedom to use his fine arboretum in Newburgh, N. Y. His superintendent, Mr. Rudolph Nocker, was also most helpful. The fine collection of native plants in this arboretum is especially outstanding.

It is impossible for me to thank all the people adequately who assisted me in many ways, but I do want to thank the following not only for their assistance in locating various shrubs, but especially for taking me to rare and wonderful places:

Professor Earl L. Core, Chairman, Department of Biology, West Virginia University

Mr. Frederic L. Steele, Tamworth, New Hampshire

Mr. & Mrs. H. Lincoln Foster, Falls Village, Connecticut

Mr. E. C. Childs, Norfolk, Connecticut

Mr. G. G. Nearing, Ramsey, New Jersey

Mr. Alton Miller, Mr. Ray Dubreuil, and Mr. Patrick Deming of the Connecticut Agricultural Experiment Station, Canaan, Connecticut

Mr. Stanley J. Smith, Curator of Botany, New York State Museum, Albany, New York

I also wish to thank the following for sending me fresh specimens:

Professor H. R. Totten, University of North Carolina

Mrs. Barbara McKenzie, Elgood, West Virginia

Mr. C. A. Eulett, Otway, Ohio

Dr. Gertrude E. Douglas, Feura Bush, New York

Mr. Kenton L. Chambers, Curator, Herbarium, Oregon State College

Dr. A. J. Sharp, University of Tennessee

For special assistance in technical matters, I thank:

Dr. Edward G. Voss, University of Michigan

Professor Oswald Tippo, Provost, University of Colorado, formerly Chairman, Botany Department, Yale University

Professor John R. Reeder, Botany Department, Yale University

Dr. Arthur Harmount Graves, Wallingford, Connecticut

Mr. Leonard J. Bradley, Staff Botanist, Audubon Center, Greenwich, Connecticut

Also thanks go to the following for their assistance in locating plants in botanical gardens:

Mr. L. P. Politi and Miss Bride McSweeney, New York Botanical Garden

Mr. George Kalmbacher, Brooklyn Botanic Garden

Dr. Donald Wyman, horticulturist, Arnold Arboretum, Jamaica Plain, Boston, Massachusetts

Mr. Henry Draper, Case Estate, Arnold Arboretum

CONTENTS

THORNS—TH

LEAVES—LF
Opposite—Opp

FLOWERS—FL

FRUIT—FR

TWIGS—TW
Opposite—Opp

BARK—BK
Opposite—Opp

MASTER PAGES—MP
Needle-leaved—NE
Ground Covers—GC
Vines—V

INTRODUCTION

This is an age of paradox and enigma:

It is an age of conservation; it is an age of wild spending of resources; an age of speed, but also an age of leisure.

It is an age for city dwellers, who travel and explore; an age of multitudes and noise, but solitude and silence are needed as never before.

This is an age of splendid sights, but it is an age beset with blindness; an age full of promise and opportunity, yet an age courting final disaster.

This is an age for greatness, but an age when simplicity is sometimes mistaken for smallness.

This is an age on a threshold of many vistas, but an age when few vistas are clear.

On this threshold, one of the paths is clear and simple. It is well trodden, but too often neglected. This is the path of simple pleasures that have always been there. Among them are the wonders of nature. This is the land free, to be used for joy and gladness, for freedom of the body and the mind. There is no bit of land so poor that discovery can be denied to those who seek it.

It has been my privilege to walk with a man who knew lichens and mosses, along a road ugly to the untrained eye. This man made the dullest-looking border become alive, full of exciting new possibilities, the squalid surroundings forgotten amidst such splendor. Here before one's eyes a miracle happened, wonders falling one over the other on the expanse of what had shortly before appeared to be only dirt, rocks and a few dusty plants. Here were wonders unexcelled in the most magnificent garden.

This time of excitement was all too short, a once in a lifetime experience. From it was gained an awareness that made possible a continuing appreciation of the less evident glories of nature. All that is required is a willingness to see what is there and a desire to understand what is seen. It is not a difficult step, but the step must be taken. The woods, the fields, the parks and wilderness areas of mountain and plain are there to be used. Even a park in the largest city has more on every hand than the largest museum. If on a walk, one really looks at the trees, shrubs, flowers, ferns, mosses and lichens, the numbers and variety are truly overwhelming until some order is made of this seeming chaos. If, then, one first picks out a few plants that are known, a start has been made, and with interest aroused, a truly fascinating pursuit begins, the thrill of personal discovery becoming real. The problems involved in identifying plants are sufficiently difficult to create a challenge, one as good as any puzzle, but more rewarding, as the solution is not the end.

In taking up a study of plants, it is perhaps best to begin with trees and shrubs as they are the easiest to identify. Success is necessary; bewilderment and failure are fatal. The goal of identification is recognition. To recognize means to re-know, and to see a familiar plant, especially in an unfamiliar setting, is indeed a thrill. This knowledge, once gained, is always at hand, and with plants available, one can never be bored.

Thus, although this book is designed specifically as a practical method for successfully identifying native and naturalized shrubs in the area covered, there should be a double incentive for using it: the pleasures of recognizing plants far outweighing the immediate objective of identifying them.

The method for shrub identification used in this book is based on the use of photographs of actual details of the plants involved. The section, "How To Use This Book," explains the process of using these details to determine the identity of any particular shrub. At this point, however, a brief explanation of botanical identification is given to avoid future misunderstanding or disappointment. As the world of plants, like that of people, is made up of individuals, it should be apparent that no two shrubs are ever exactly alike. There are, nevertheless, certain characteristics possessed by every shrub which indicate its relationship to other shrubs. For example every Lilac has a number of features that are similar to those of all Lilacs and which distinguish it from other shrubs. It is by a deliberate and directed use of these details that identification is accomplished, not only by the botanist but by anyone attempting to do so.

It sometimes happens that some of the details vary among plants of the same kind, which is confusing to the novice; but despite these apparent discrepancies, all plants do exhibit a pattern and consistency of detail which makes classification always possible. Only the matter of degree is the stumbling block. For example, great variations may occasionally be encountered, but, as there is always a cause, logical reasons can usually be found to account for them. All plants have evolved over long periods of time, and when the environment has remained relatively constant long enough, stable forms are established. When the environment has not remained constant, for one reason or another, plants may not achieve (or retain) a stable form. This condition is particularly evident among certain kinds of plants that have been directly affected by extensive clearing of land in this country. These exceptions to the standard form are indicated, but few, if any, specific names are given for such plants in this book. It should be understood that it is not necessary to attempt exact identification of a shrub under these conditions, but rather to place it in the small group of related plants to which it is found to belong. This is almost always a practical and correct solution, infinitely more correct than trying to force an exact (named) identity on a plant when none exists. The great majority of shrubs are stable enough so that precise identification is usually possible. Final identification in this book, except for the unstable forms mentioned above, means reducing the possibilities to the correct species, such as Mountain Laurel or Flowering Dogwood. The botanist carries the process further into varieties (*Pink* Flowering Dogwood, for example), hybrids (crosses between species) and specialized forms, but this is not generally necessary for practical purposes, or, as in the case of the Pink Flowering Dogwood, the variation from the standard is obvious.

The shrubs included in this book grow within an area bounded roughly by the eastern half of the United States and Canada. As this book is not a complete manual, certain plants were omitted so that unnecessary numbers would not interfere with the main purpose. Records of all woody plants known to be growing in the area were studied and compared with reports obtained from over twenty states. From this study, a pattern of distribution and frequency emerged for each shrub, which proved to be a valuable guide in determining which shrubs to include. The omissions fall mostly into the following categories:

(1) Shrubs from other regions that overlap small sections of the area covered by this book.

(2) Strictly southern or subtropical plants.

(3) Foreign shrubs which, although planted here, have not become naturalized to any extent in this country. (A number of foreign shrubs have become established here and now appear to be almost "native." These are included.)

(4) Hybrids, varieties and unusual forms.

(5) Some shrubs belonging to genera whose species are very numerous. It was felt in these cases that a number of the most important and representative species would be sufficient, for once the genus is known, identification of any additional species can be made with the help of a botanical manual, available at most libraries, or, usually, by consulting a list describing the local flora. The Hawthorns have hundreds of species, but the specialists (people who may spend years on this genus alone) never agree as to just how many there are or when to call a particular plant a real species rather than only a variety of another. In this case, only the genus (Hawthorn) is given here, with the warning that naming of species is better left to others.

Both popular and botanical names are given for each shrub, the botanical ones in accordance with Alfred Rehder's *Manual of Cultivated Trees and Shrubs,* Macmillan, second edition. There are two exceptions to this: the Clubmosses (*Lycopodium*) and Japanese Knotweed (*Polygonum cuspidatum*). These are not included in Rehder's manual, and the botanical names follow the eighth edition of *Gray's Manual of Botany,* by M. L. Fernald.

HOW TO USE THIS BOOK

The objective in any but the most casual system of identification is the classi-fication as well as the naming of something. In one way or another, the method for doing this is based on the elimination of possibilities. This usually involves the reduction of large numbers of such possibilities to fewer and finally to one. Plant identification is based on plant relationships. Botanical names (and some common or popular names) are also based on these relationships. For practical purposes, there are two terms to keep in mind: (1) genus and (2) species. *Genus* is applied to a closely related group of plants, such as the Dogwoods or the Honeysuckles. The *species* is the particular kind of Dogwood or Honeysuckle. The botanist places the genus name first, with a capital letter, followed by a specific epithet in small letters, thus: *Cornus florida*—Flowering Dogwood (*Cornus*—Dogwood; *florida*—Flower-ing). Botanical nomenclature has several advantages: It is understood in all countries, is based on natural relationships of plants and is consistent in form. Common names, on the other hand, were not originated with these features in mind. However, as botanical names are unfamiliar to many people and appear somewhat as mumbo-jumbo to the uninitiated, common names have been used throughout this book. The disadvantage of doing this is, of course, that the particular name selected is not always the best known in all sections of the country. To overcome this, alter-native popular names (together with the botanical *one* for each shrub) are given in the main section of the book devoted to information concerning individual plants, with an apology made here for the omission of someone's favorite.

The method used in this book to identify shrubs utilizes photographs of specific parts of each shrub. By placing similar details, such as leaves, together in separate groups, the groups are quickly found, and the differences between differentiating details within each group are easily noted, thus eliminating all but a few possibilities. Finally, by comparing these few with a few possibilities within an entirely different group (or groups), all shrubs but the correct one are eliminated.

There are five main parts (or details) of a shrub to look for: Leaves, Flowers, Fruit, Twigs and Bark. Each of these has a number of smaller details which are pointed out in the book, all of which can easily be seen in the photographs. All of the five major details may not be available at any one season of the year, but usually a combination of any two or three is sufficient for practical purposes. In the very few cases when this is not so, it may be necessary to wait for an additional one. This should not be a source of discouragement when one realizes that a botanist might have to wait also. Plants are not machine-made products, but living, changing things, and one must not expect that each individual shrub will be exactly like every other. One can expect, however, that the individuals grouped together under one specific name will be sufficiently consistent in the total combination of their details to make identification reasonably certain.

A shrub is usually defined as a relatively small, woody plant with several stems (instead of a single trunk as with most trees). It might be simpler to think of a shrub as any woody plant that is not a tree, as the terms "tree" and "shrub" are actually used only as a convenient means of dividing woody plants into two easily recognizable groups. For further convenience in reducing large numbers to smaller, shrubs have been divided here into four readily distinguishable groups: (1) Broad-leaved upright shrubs, (2) Needle-leaved shrubs of all kinds, (3) Broad-leaved ground covers, and (4) Vines. The plants in each of these groups are in separate sections of Part II of this book, called the Master Pages. The broad-leaved *upright* shrubs constitute by far the largest and therefore the most difficult group to identify, and for this reason the first part of this book is in the form of Pictorial Keys which are designed for the identification of the *genus* of these plants. The last step, *species* identification, is accomplished in the second part of the book. When a plant belongs to one of the last three groups above, reference should be made directly to the appropriate section of the Master Pages. As there are relatively few of these plants, no special Key has been given for them. They are all distinctive and can be found quickly within the few pages of the sections involved. Before using this book, the Introductions to these three sections should be consulted to see just what types of plants are included in each of them.

There are six Keys for genus identification of the broad-leaved *upright* shrubs. In addition to a Key for each of the five major details mentioned previously, a Key for thorns is given first, as it is a quick guide when thorns are present. The other five Keys are the basic ones, however. Note that the contents page of this book is indexed at the edge to correspond to similarly placed markings on the outer edges of the pages of the book, making it easy to find a particular Key or Master Page section. Each Key, except for the Thorn one, has a short Introduction explaining how it is arranged. These should be read before using the Keys for the first time; thereafter they will be found useful as checklists of things to look for.

When using the Keys, start with the first one and look for thorns on the actual plant, then continue through the Keys in the order in which they are presented, until satisfied that the genus has been correctly determined. Do not start with Key #6, Bark, for example, as this Key is designed only to verify identification made from previous Keys. Also, do not dismiss leaves or fruit in winter, as they may hang on for much of the year. Twigs, although primarily a winter characteristic, are often very helpful in summer, the buds often being well developed before the leaves fall off. It will become apparent with actual use that a very good idea of identity is obtained from the first applicable Key; however, several Keys should be used to avoid jumping to faulty conclusions, and will be found essential for the plants that are difficult to identify. If, for example, it is difficult to decide between the pictures of three different twigs, turn to the Bark Key, and the chances are good that the bark of the actual plant will be like only one of the three possible plants selected in the Twig Key. Emphasis is again made that final identification should never be based on one detail alone. Any specific detail of an individual plant of any one species may vary considerably from the average for the species, but the combination of all available details will be found to be indicative for that plant and species. (In cases when this is not found to be so, the plants will almost certainly prove to be varieties, hybrids or other unusual forms.)

A very important feature to keep in mind is the Alternate and Opposite characteristics of shrubs. The leaves, buds and leaf-scars (the places marking where leaves separated from the twigs when they dropped off in the fall) of every plant are arranged either *alternately* along the stems, or in *opposite* pairs (occasionally in whorls of three to four around the stem at definite points called nodes). This is a consistent characteristic of almost all woody plants. The Leaf, Twig and Bark Keys have been subdivided into Alternate and Opposite sections, and the first step in identifying a shrub is to notice whether it is alternate or opposite. Two shrubs that are exceptions to the rule commonly have both alternate and opposite leaves and buds on the same plant and are represented in both sections of the Keys. A few shrubs that are actually alternate produce leaves in clusters at the ends of the twigs, giving an opposite impression. These will be found in both sections of the Leaf and Bark Keys, but only in the Alternate section of the Twig Key, as their buds are obviously alternate.

Under each picture in the Keys is a name and a number. When going from one Key to the next, always look for the correct picture and not for the name selected in the previous Key, as several different pictures are frequently given for the same genus. Only when the picture corresponds to the actual detail should the name correspond to the one previously selected, thus verifying identification. When the genus has been determined, turn to the Master Page indicated by the number shown next to the correct name. The Master Pages bring together all the details of the individual shrubs. Whenever a particular genus, such as Dogwood, includes two or more species, the corresponding details of the different species are shown together, so that final (species) identification is made by contrasting these different details. If the genus of a particular shrub is already known, it is not necessary to use the Keys, and by consulting the index in the back of the book, the correct Master Page can be turned to directly for species identification.

The pictures in this book are all actual size except in a few cases, and then the scale is indicated. Details that are to be compared are always in the same scale.

A small hand lens is useful and interesting to use, but is not necessary except in cases where very small details must be clearly seen, and this mostly to distinguish between very similar varieties or forms of the same species, or for other specialized work.

PART I

PICTORIAL KEYS

BROAD-LEAVED UPRIGHT SHRUBS
(all actual size except when noted)

CONTENTS: Key #1 Thorns, Prickles, Bristles and Similar Growth

Key #2—Leaves

Key #3—Flowers

Key #4—Fruit

Key #5—Twigs

Key #6—Bark

KEY #1 THORNS, PRICKLES, BRISTLES
and Similar Growth

The leaves, buds and twigs of the three plants in this top row are found opposite one another; all other thorny plants in this book produce alternate growth.

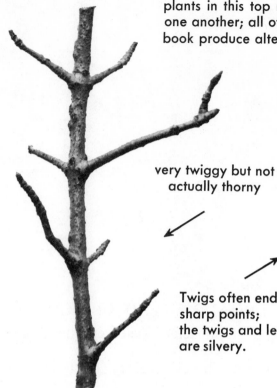

very twiggy but not actually thorny

Twigs often end in sharp points; the twigs and leaves are silvery.

Buckthorn 84

Viburnum 136

Buffalo-berry 9

No actual leaves are produced, only leaf-stalks, which are sharp and thorn-like.

evergreen

Gorse 67B

Firethorn 42

Plum 54

Oleaster 96

Hawthorn 43

Sarsaparilla 92

small stems and twigs round;
green, red or maroon

buds usually small; shiny
red (or green)

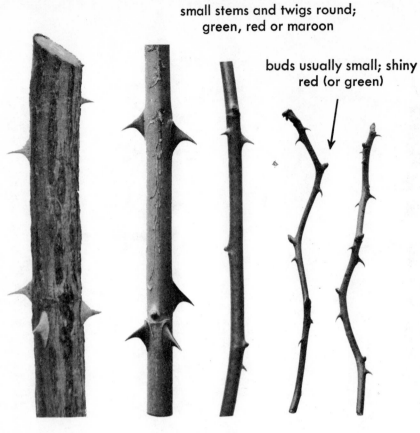

rust
color-
bud

Rose 57

Prickly Ash 70

stems hardly woody, with very large pith;
often angled (not round); green, red or maroon

Remnants of leaf-stalks (below bud)
often persist on stems.

**Bramble 46
(Blackberry)**

Stems are actually red, but almost always covered by a heavy bloom that makes them look white or pinkish.

Bramble 46
(Raspberry)

small stems and twigs round; green, red or maroon; buds usually small, shiny red (or green)

small stems and twigs round; green, red or maroon

Rose 57

Rose 57

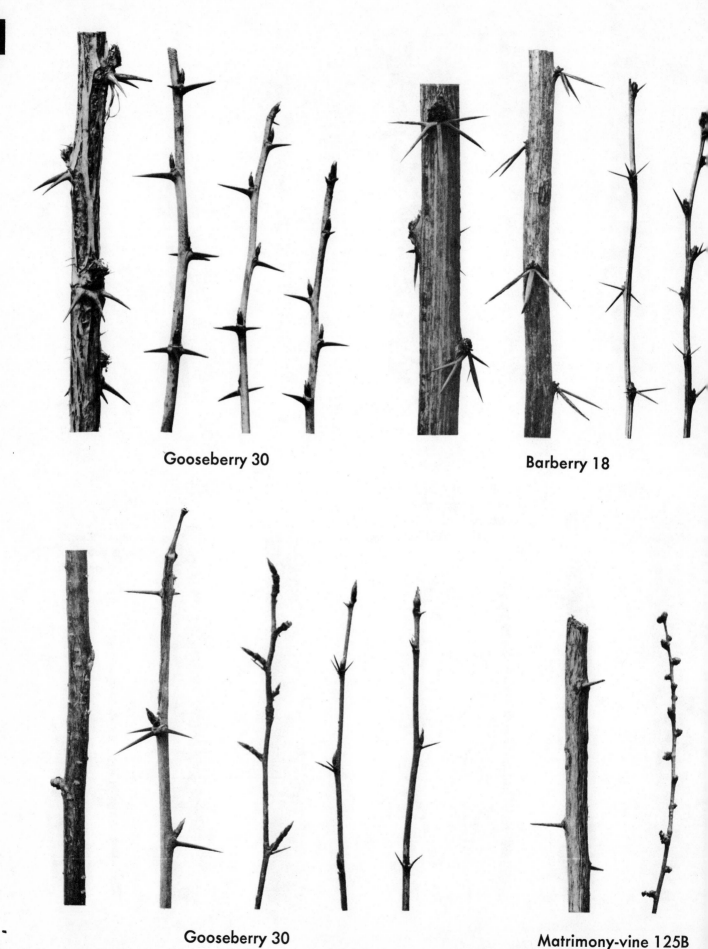

Gooseberry 30

Barberry 18

Gooseberry 30

Matrimony-vine 125B

rough bark;
dark gray
or brown

small stems and twigs round; green,
maroon or red; buds usually small,
shiny red (or green)

stems often angled (not round); green,
maroon or red; remnants of leaf-stalks
often persist on stems into winter

Rose 57

Bramble 46
(Blackberry)

Barberry 18

Gooseberry 30

Thorns and Bristles

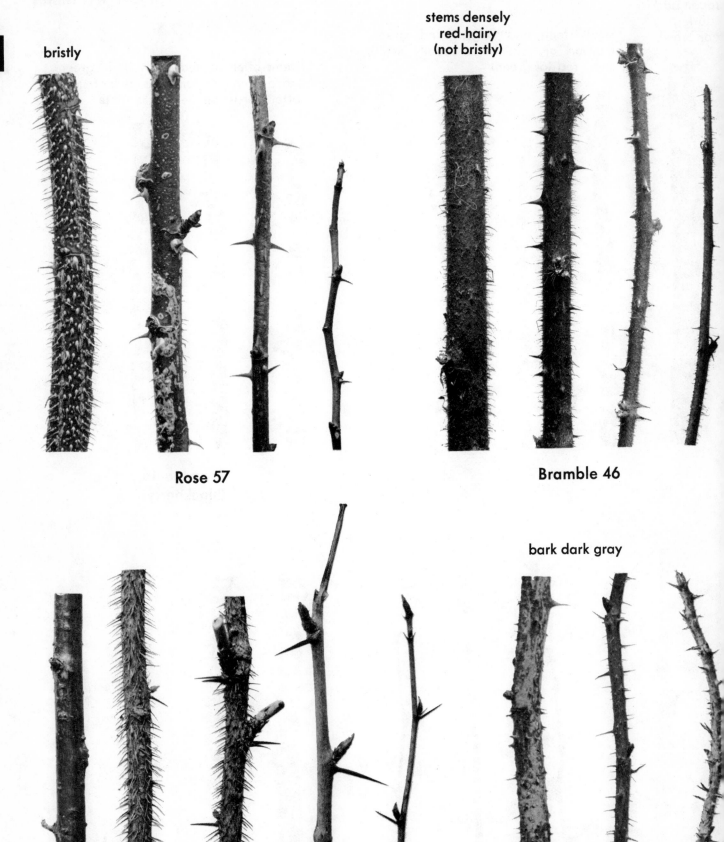

bristly

stems densely
red-hairy
(not bristly)

Rose 57

Bramble 46

bark dark gray

Gooseberry 30

Rose 57

stems golden brown (or red),
with or without bloom

twigs often bristly

small stems and twigs
green, red or maroon

buds shiny
red (or green)

**Bramble 46
(Raspberry)**

Rose 57

bark of larger
stems dark gray

bristles
rust colored

stems, twigs and
bristles golden brown

Bristly Locust 69

Currant 30

Sarsaparilla 92

Key #2—LEAVES

This Key is arranged as follows:

I ALTERNATE LEAVES, pp. 1-25

    A Simple Leaves—one main leaf blade on leaf-stalk (petiole)

        (1) With teeth evident along edges of leaf, pp. 1-11

        (2) Without teeth (entire) or with indistinct teeth, pp. 12-19

    B Compound Leaves—with more than one leaflet on leaf-stalk, pp. 20-25

II OPPOSITE LEAVES, pp. 26-36

    A Compound Leaves, p. 26

    B Simple Leaves

        (1) With teeth, pp. 27-30

        (2) Without teeth or with indistinct teeth, pp. 31-36

Note: Some shrubs which actually have alternate characteristics often produce leaves in clusters at the ends of the twigs, giving an opposite effect. These leaves will be found in both the alternate and opposite sections. Two shrubs, Common Buckthorn and Fringe Tree, although predominantly opposite, often have some alternate leaves and these appear in both sections. Shrubs which often have different types of leaves are represented on more than one page.

The simple leaves in each group are arranged with the largest leaves first and with the very small or very narrow ones last.

In comparing actual leaves with the photographs, look particularly for the following five characteristics:

(1) EDGE OF LEAF: if with teeth, note differences in size, shape, and direction (whether pointing outward or forward toward the end of the leaf); also whether singly or doubly toothed (small teeth on larger ones); also whether the teeth are all the same size and evenly or unevenly spaced. Sometimes there are no regular teeth but the leaf is wavy-edged. A few shrubs have leaves both with and without teeth, in which case they are shown in two or more places in the Keys.

(2) TIP OF LEAF: long-tapered, short-pointed, sharp-pointed, blunt or rounded.

(3) BASE OF LEAF: tapered, rounded, straight across or heart-shaped

(4) LEAF-STALK (petiole): long, short, wide or narrow

(5) VEINS: arrangement—large (primary) veins branching off along whole length of midrib (pinnately), or starting mostly at base of leaf from approximately one point (palmately); also note whether the main veins are straight and go directly to the edges of the leaf, or whether they branch before reaching the edges; also note if veins curve upward and whether they are prominent or indistinct.

A brief examination of this Key will indicate that leaves which appear to be superficially alike will almost always have at least one of the above characteristics markedly different. Be sure when using this Key that the differences noted are clear-cut. For example, if the leaves of a particular shrub have very short leaf-stalks it would be safe to assume that leaves of all shrubs of the same kind would also have short leaf-stalks and would be found so in the Key pictures. Do not, on the other hand, try to distinguish between leaves based on leaf-stalks in the middle range. Do not base final identification on leaf size, as leaves vary greatly in this respect. In fact, it should be noted that, in collecting the leaves for this Key, a deliberate effort was made to show all similar leaves in the same size so that variations of size would not influence comparisons. The five details listed above are the ones to concentrate on after the general average size is determined. Occasionally, a note is given of other characteristics as, for example, if leaves are distinctly whitened beneath or covered with evident dots, this is indicated. Otherwise, it can be assumed that the five points are sufficient to distinguish between the leaves shown.

I. ALTERNATE LEAVES

sticky (glandular),
hairy leaf-stalk

Bramble 46
(Raspberry)

Currant 30

Currant 30

Currant 30

Currant 30

Currant 30

Gooseberry 30

Gooseberry 30

Gooseberry 30

Hawthorn 43

Ninebark 27

LF 3

Groundsel-bush 155

Oak 14

Oak 14

Chestnut 13
(Chinquapin)

Hazelnut 10

Rose of Sharon 90

Witch-Hazel 38

LF
4

Silverbell 128

Alder 6

Holly 78

Alder 6

Alternate Leaves

Hazelnut 10

Alder 6

Alder 6

Hornbeam 12

Shadbush 44

Spirea 40

Birch 9

34

LF 6

Plum 54   Buckthorn 84   Plum 54

Clethra 102   Holly 78

Alternate Leaves

New Jersey Tea 45A

Virginia-willow 23B

Leucothoë 118

usually
opposite,
occasionally
partly alternate

Buckthorn 84

Alder 6

Holly 78

Buckthorn 84

Holly 78

Leucothoë 118

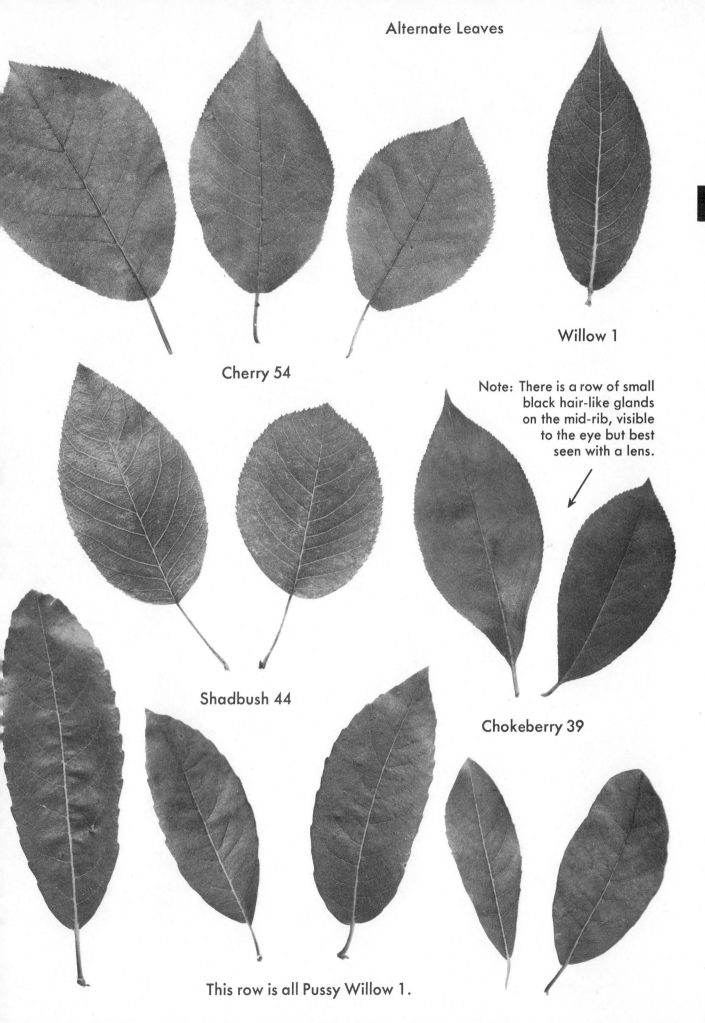

Alternate Leaves

Willow 1

Cherry 54

Note: There is a row of small black hair-like glands on the mid-rib, visible to the eye but best seen with a lens.

Shadbush 44

Chokeberry 39

This row is all Pussy Willow 1.

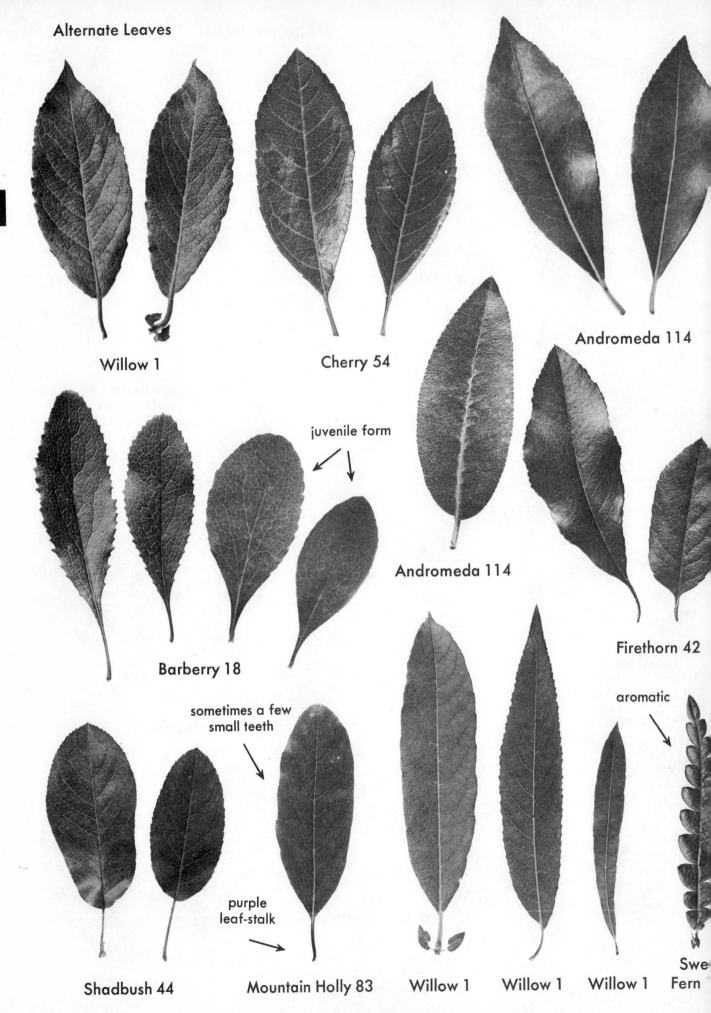

Alternate Leaves

Willow 1

Cherry 54

Andromeda 114

juvenile form

Barberry 18

Andromeda 114

Firethorn 42

sometimes a few
small teeth

aromatic

purple
leaf-stalk

Shadbush 44

Mountain Holly 83

Willow 1

Willow 1

Willow 1

Swe
Fern

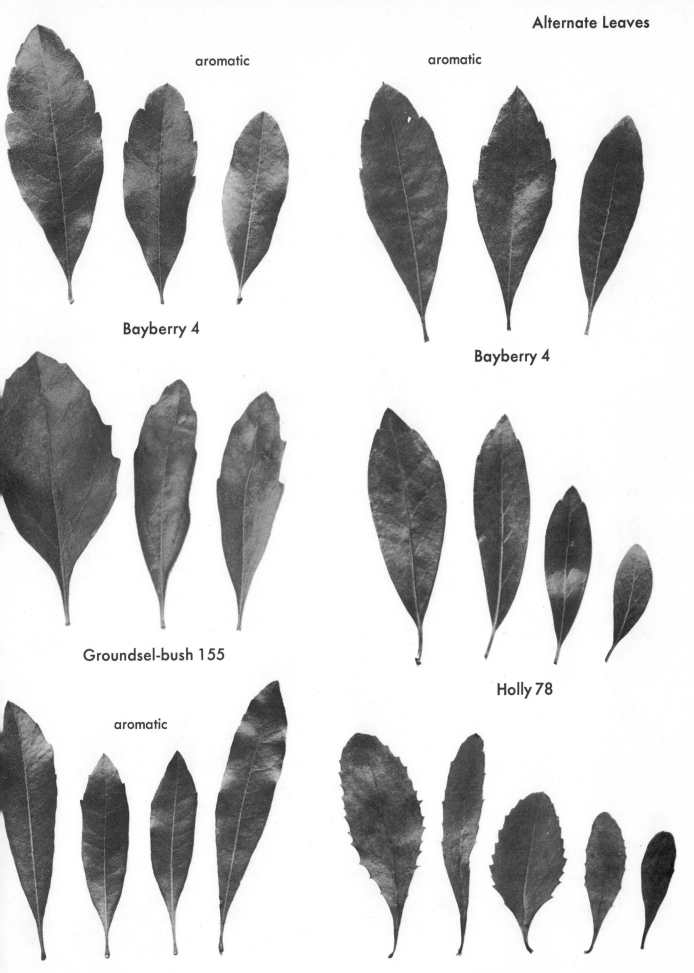

aromatic

aromatic

Bayberry 4

Bayberry 4

Groundsel-bush 155

Holly 78

aromatic

Wax-myrtle 4

Barberry 18

Alternate Leaves

aromatic

Sweet Gale 4

Spirea 40

Hawthorn 43

upper leaves usually
alternate; lower
ones opposite

Spirea 40

Birch 9

Marsh-elder 156

leaves
covered
with many
small dots

Leatherleaf 116B

pale, often rusty-woolly,
beneath

Spirea 40

Barberry 18

Blueberry 119
(Bilberry)

Huckleberry 126

Knotweed 17B

Smoke-tree 72

Redbud 68

Rhododendron 104

Rhododendron 104

Alternate Leaves without Teeth

mostly opposite, but occasionally
partly alternate

aromatic

Fringe-tree 129

Spice-bush 21

small teeth
not always
evident

veins particularly
conspicuous early
in the season

Leucothoë 118

Buffalo-nut 16

Alternate Leaves without Teeth

usually with teeth, but edges often curl down, hiding them

Smoke-tree 72

Alder 6

Dogwood 98
(Alternate-leaved)

Buckthorn 84

Lyonia 117

Leatherwood 91A

43

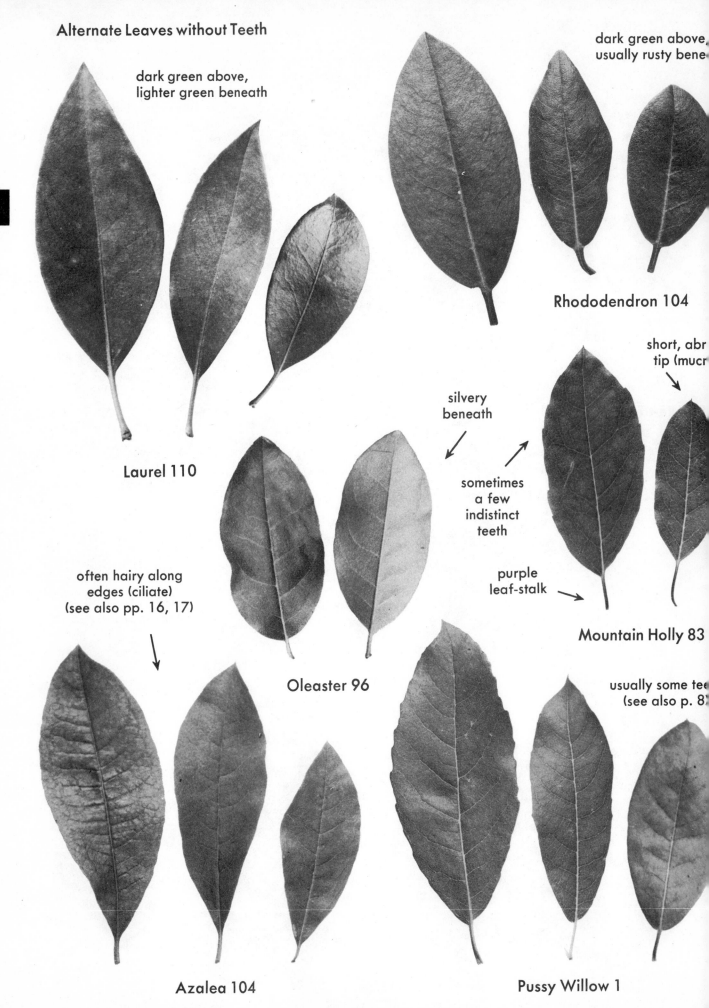

Alternate Leaves without Teeth

dark green above,
lighter green beneath

dark green above,
usually rusty bene

Rhododendron 104

Laurel 110

silvery
beneath

short, abr
tip (mucr

sometimes
a few
indistinct
teeth

often hairy along
edges (ciliate)
(see also pp. 16, 17)

purple
leaf-stalk

Mountain Holly 83

Oleaster 96

usually some tee
(see also p. 8

Azalea 104

Pussy Willow 1

often hairy along edges (ciliate) (see also pp. 15, 17)

Lyonia 117

Azalea 104

rough (to touch) along edges, but no regular teeth
(see also p. 17)

(see also p. 14)

Leatherwood 91A

Lyonia 117

This row is all Blueberry 119.

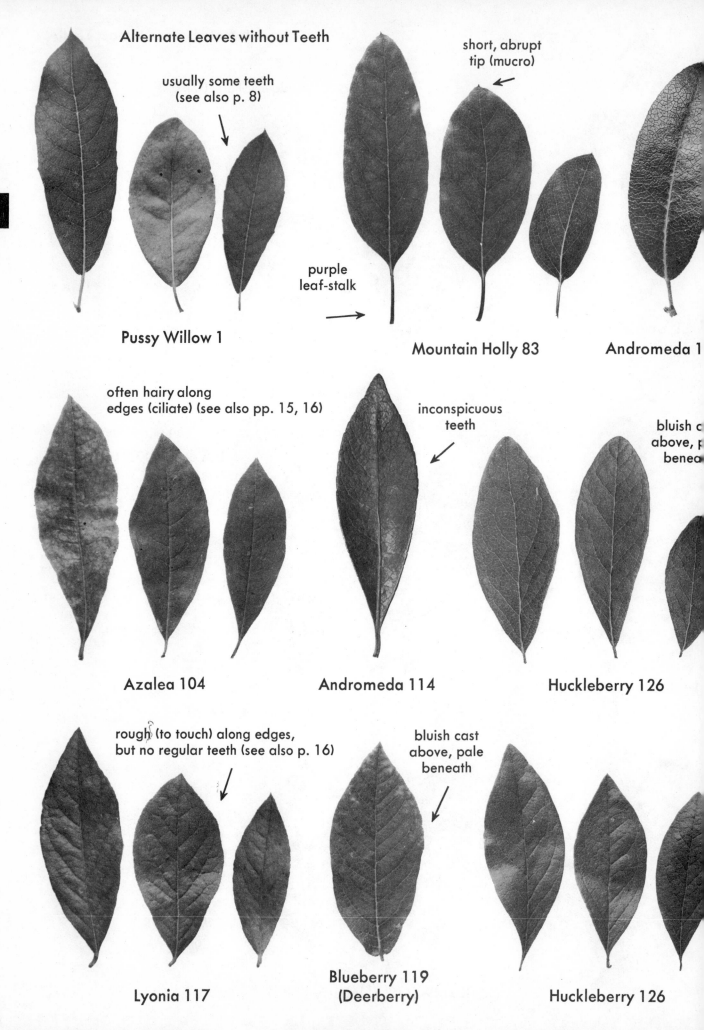

Alternate Leaves without Teeth

usually some teeth
(see also p. 8)

short, abrupt
tip (mucro)

purple
leaf-stalk

Pussy Willow 1

Mountain Holly 83

Andromeda 1

often hairy along
edges (ciliate) (see also pp. 15, 16)

inconspicuous
teeth

bluish c
above, p
benea

Azalea 104

Andromeda 114

Huckleberry 126

rough (to touch) along edges,
but no regular teeth (see also p. 16)

bluish cast
above, pale
beneath

Lyonia 117

Blueberry 119
(Deerberry)

Huckleberry 126

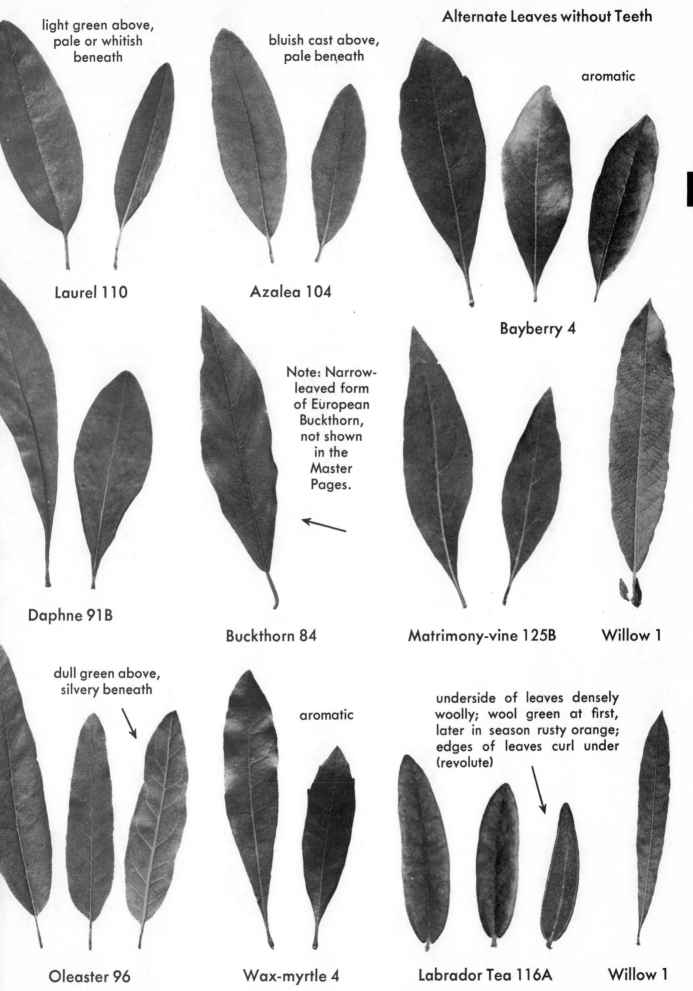

light green above, pale or whitish beneath

bluish cast above, pale beneath

aromatic

Laurel 110

Azalea 104

Bayberry 4

Daphne 91B

Note: Narrow-leaved form of European Buckthorn, not shown in the Master Pages.

Buckthorn 84

Matrimony-vine 125B

Willow 1

dull green above, silvery beneath

aromatic

underside of leaves densely woolly; wool green at first, later in season rusty orange; edges of leaves curl under (revolute)

Oleaster 96

Wax-myrtle 4

Labrador Tea 116A

Willow 1

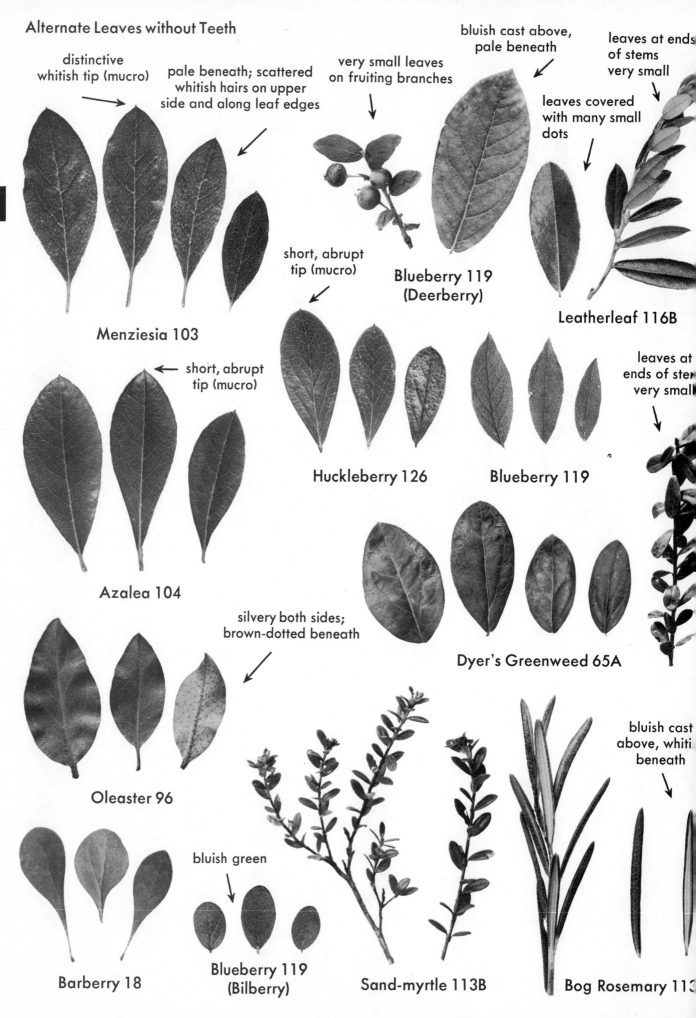

Alternate Leaves without Teeth

distinctive whitish tip (mucro)

pale beneath; scattered whitish hairs on upper side and along leaf edges

very small leaves on fruiting branches

bluish cast above, pale beneath

leaves at ends of stems very small

leaves covered with many small dots

Menziesia 103

short, abrupt tip (mucro)

Blueberry 119 (Deerberry)

Leatherleaf 116B

short, abrupt tip (mucro)

leaves at ends of ste... very small

Huckleberry 126

Blueberry 119

Azalea 104

silvery both sides; brown-dotted beneath

Dyer's Greenweed 65A

bluish cast above, whiti... beneath

Oleaster 96

bluish green

Barberry 18

Blueberry 119 (Bilberry)

Sand-myrtle 113B

Bog Rosemary 113

48

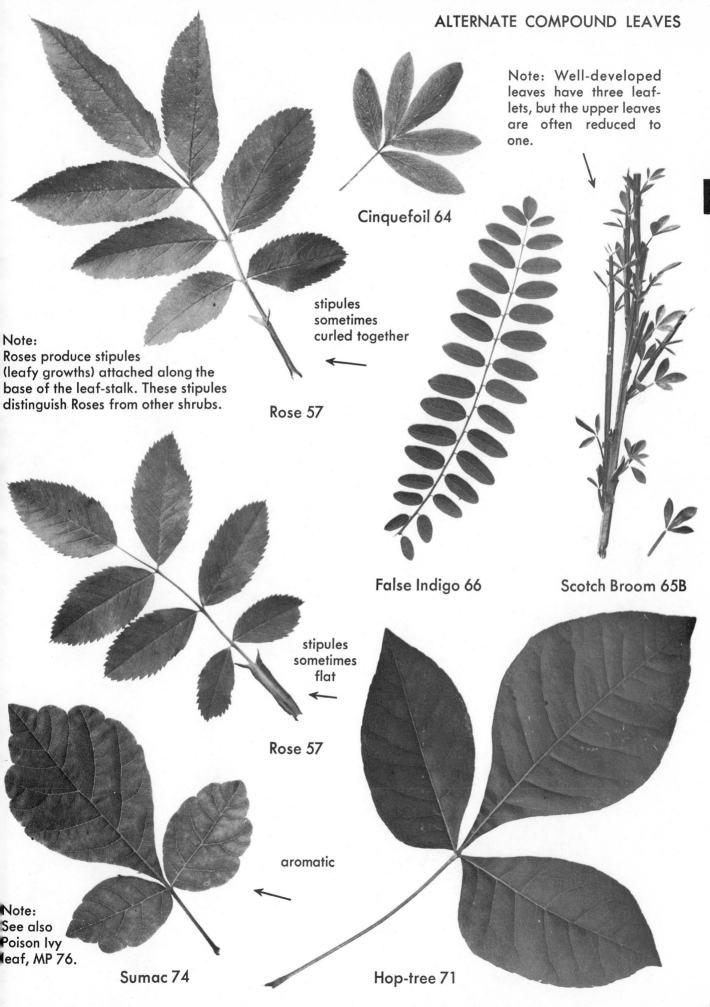

Note: Well-developed leaves have three leaf-lets, but the upper leaves are often reduced to one.

Cinquefoil 64

stipules sometimes curled together

Note:
Roses produce stipules (leafy growths) attached along the base of the leaf-stalk. These stipules distinguish Roses from other shrubs.

Rose 57

False Indigo 66

Scotch Broom 65B

stipules sometimes flat

Rose 57

aromatic

Note:
See also
Poison Ivy
leaf, MP 76.

Sumac 74

Hop-tree 71

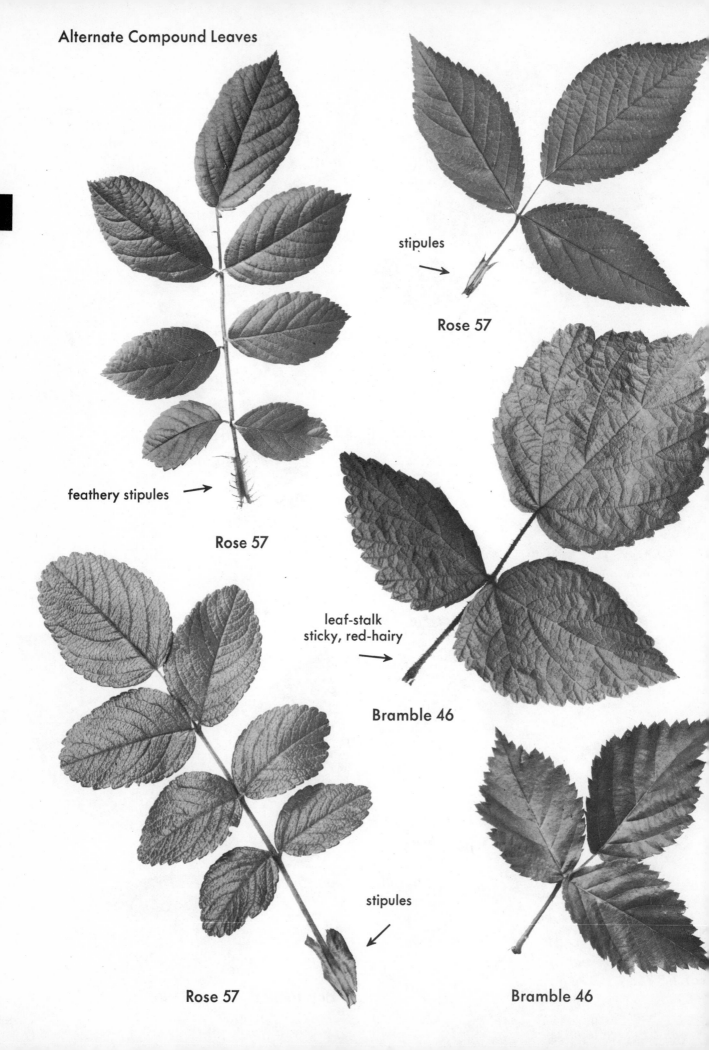

stipules

Rose 57

feathery stipules

Rose 57

leaf-stalk
sticky, red-hairy

Bramble 46

stipules

Rose 57

Bramble 46

LF
22

This page is all Brambles 46
(Blackberry and Raspberry).

Alternate Compound Leaves

This page ½
actual size

Note wings on main
leaf-stalk.

no wings
no teeth

small teeth along
edges of leaflets

Dwarf Sumac 74
(non-poisonous)

Poison Sumac 74
(poisonous; worse
than Poison Ivy
for many people)

Prickly Ash 70

Bristly Locust 69

False Indigo 66

Sarsaparilla 92

Shrub Yellow-root 25

single teeth
along edges
of leaflets

doubly serrate
(small teeth on
large ones)

sometimes
very large

Sumac 74

False Spirea 24

½ actual size

This is all one compound
leaf; these leaves
are often 3 feet
in length.

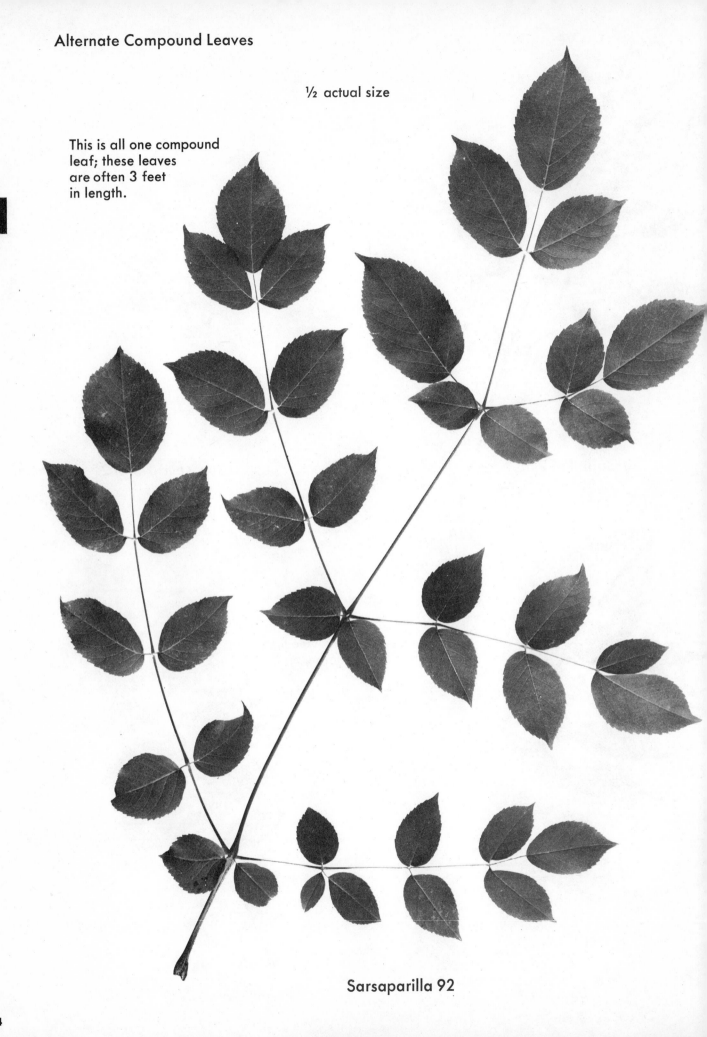

Sarsaparilla 92

. OPPOSITE LEAVES

(all are actual size)

Elderberry 134

Bladdernut 87

Forsythia 131

juvenile form
of leaf often
found on new
shoots or young
growth

All leaves on this page are Viburnum 136.

Viburnum 136

Jetbead 45B

Viburnum 136

Mock-orange 22

Hydrangea 28

LF
29
Opp

red
leaf-stalk ←

reddish
or pinkish
leaf-stalk →

Hydrangea 28   **Viburnum 136**   **Viburnum 136**

often
wavy-edged
leaf-stalk →

Bush-honeysuckle 154   **Euonymus 80**   **Viburnum 136**   **Euonymus 80**

leaf wavy-edged
with or without small
rounded teeth;
margin often
curls under

Euonymus 80

Viburnum 136

These two mostly without teeth
see also   see also
p. 35   p. 36

Deutzia 26

Buckthorn 84

Coralberry 145

Snowberry 145

Euonymus 80

lower leaves
opposite, upper
ones largely
alternate

mostly alternate,
rarely partly
opposite

Viburnum 136
e: This is a narrow-
ved form of Nanny-
ry, not shown in the
ster Pages.

Bush-honeysuckle 154   Forsythia 131   Marsh-elder 156   Andromeda 114

LF
30
Opp

59

OPPOSITE LEAVES WITHOUT TEETH (entir
or with Indistinct Teeth

LF 31 Opp

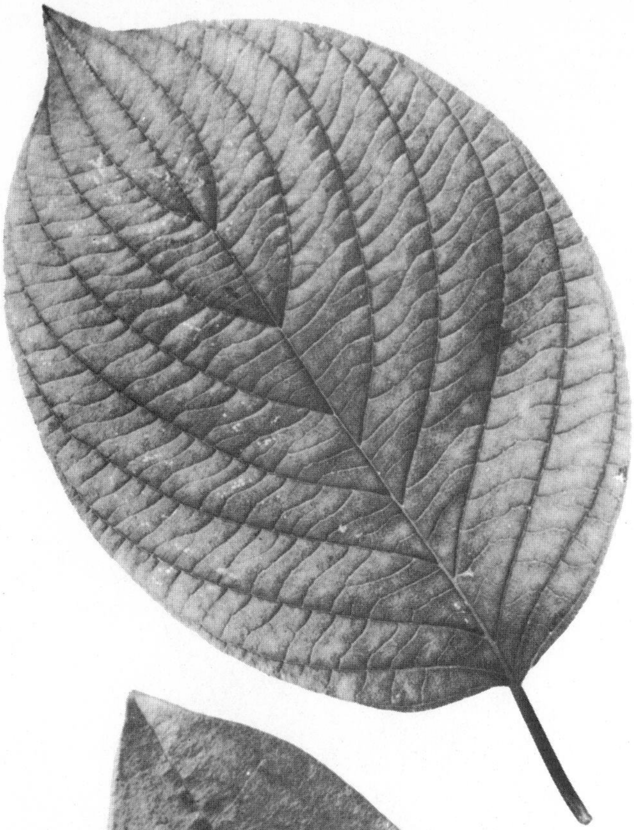

Dogwood 98

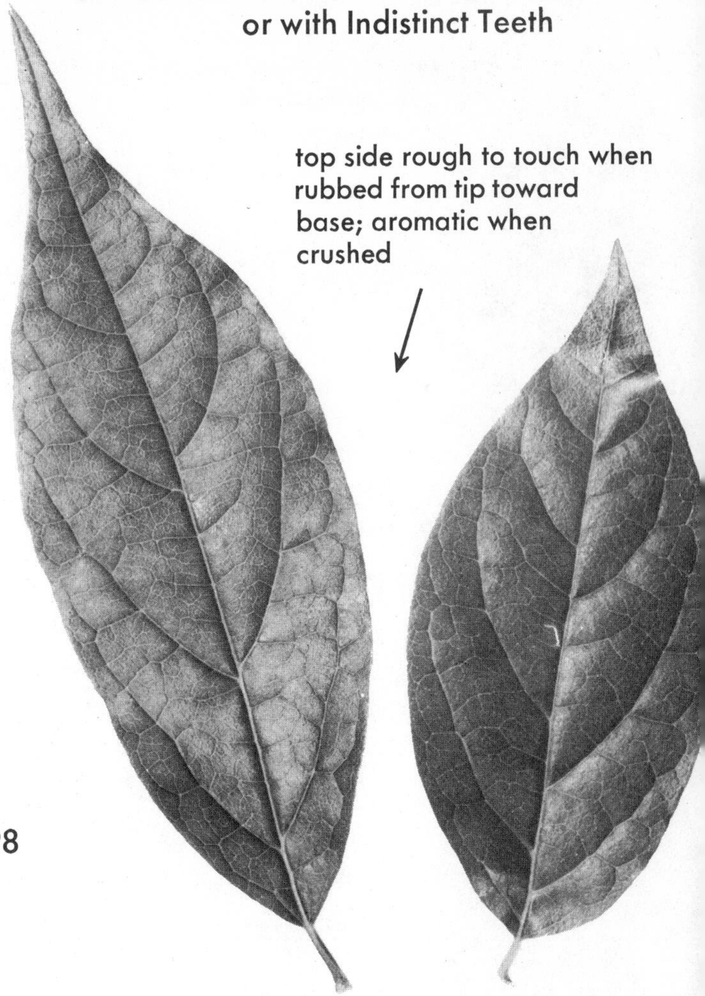
top side rough to touch when rubbed from tip toward base; aromatic when crushed

Sweet Shrub 20

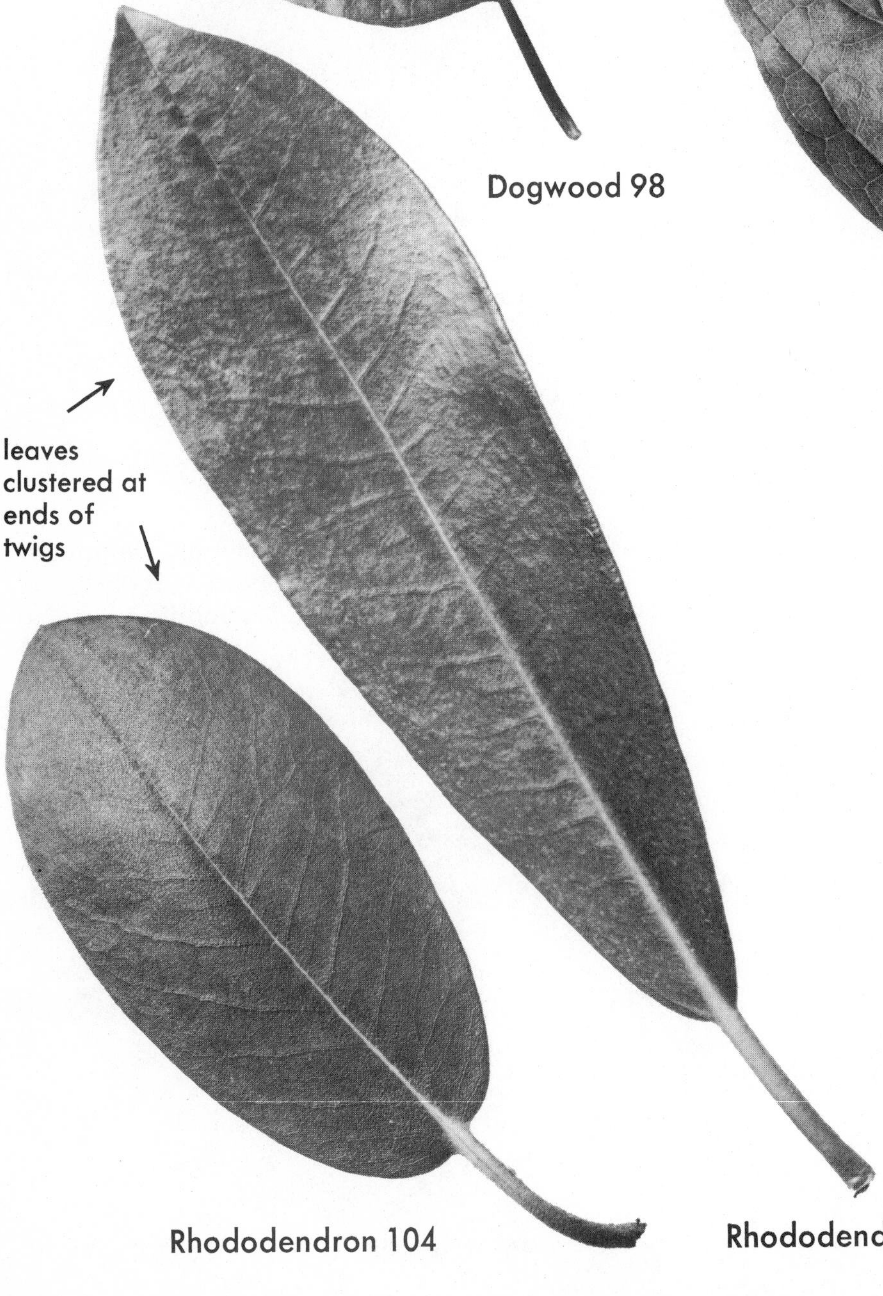
leaves clustered at ends of twigs

Rhododendron 104

Rhododendron 104

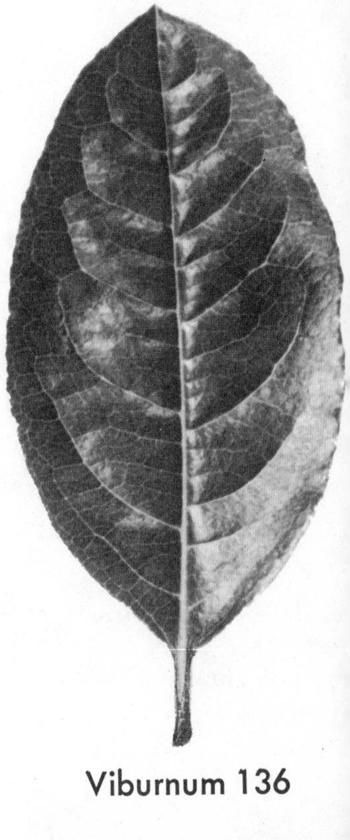

Viburnum 136

mostly in whorls of
three or four around stem
at each joint

Buttonbush 144

Dogwood 98

Dogwood 98

Fringe-tree 129

Opposite Leaves without Teeth

This row is all Dogwood 98.

Lilac 130

Honeysuckle 146

Mock-orange 22

Honeysuckle 146

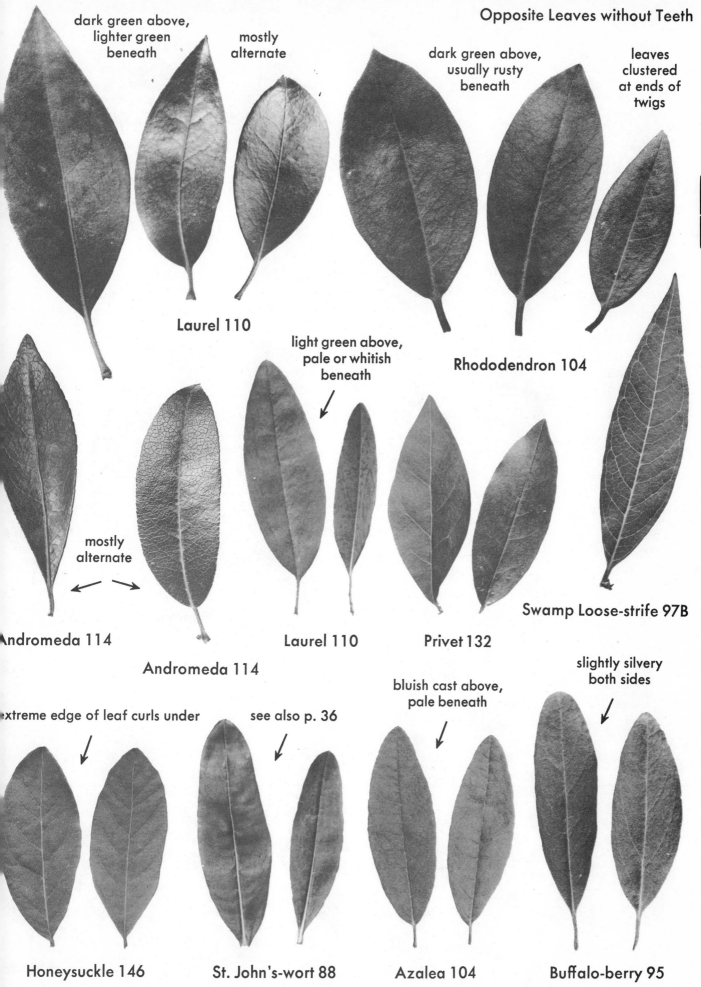

dark green above, lighter green beneath

mostly alternate

dark green above, usually rusty beneath

leaves clustered at ends of twigs

Laurel 110

Rhododendron 104

light green above, pale or whitish beneath

mostly alternate

Andromeda 114

Andromeda 114

Laurel 110

Privet 132

Swamp Loose-strife 97B

extreme edge of leaf curls under

see also p. 36

bluish cast above, pale beneath

slightly silvery both sides

Honeysuckle 146

St. John's-wort 88

Azalea 104

Buffalo-berry 95

LF
35
Opp

Snowberry 145

Honeysuckle 146

Honeysuckle 146

Honeysuckle 146

leaves alternate,
but clustered at ends of twigs;
often hairy along edges

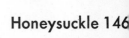

Azalea 104

Honeysuckle 146

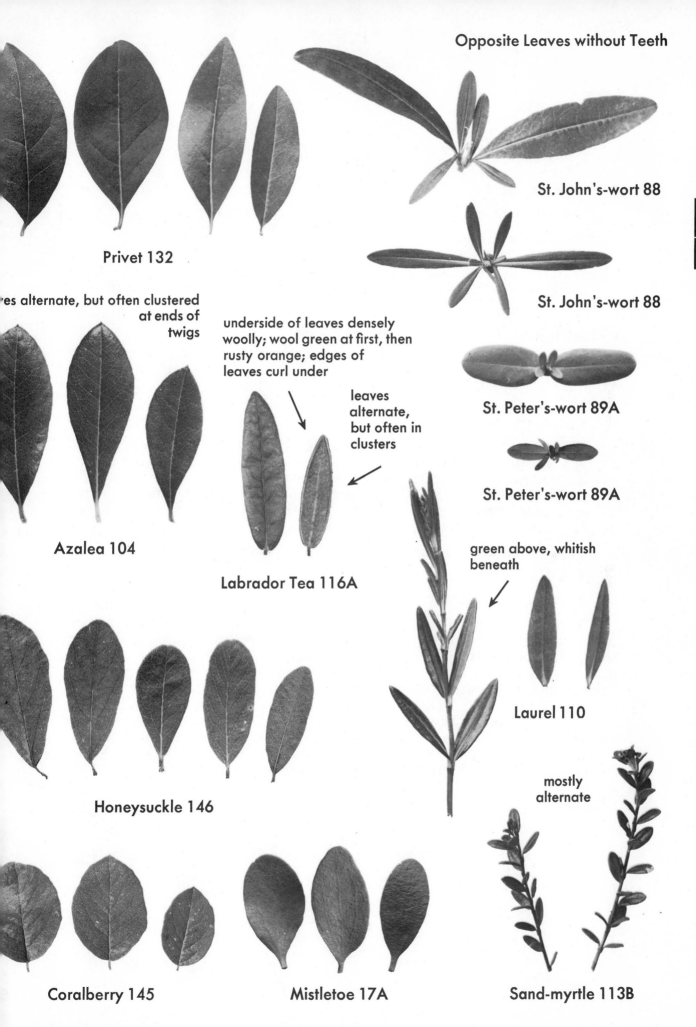

St. John's-wort 88

St. John's-wort 88

Privet 132

es alternate, but often clustered at ends of twigs

underside of leaves densely woolly; wool green at first, then rusty orange; edges of leaves curl under

leaves alternate, but often in clusters

St. Peter's-wort 89A

St. Peter's-wort 89A

Azalea 104

Labrador Tea 116A

green above, whitish beneath

Laurel 110

Honeysuckle 146

mostly alternate

Coralberry 145

Mistletoe 17A

Sand-myrtle 113B

65

Key #3—FLOWERS

This Key is arranged as follows:

Note: Within each of the above groups the earliest blooming flowers are placed first, the latest blooming ones last. Flower colors are noted when distinctive.

The divisions and arrangement of this Key were determined solely to make finding any particular flower easier; the headings were chosen for the same purpose and are not intended to be technically descriptive nor scientifically correct. The botanical nomenclature for floral parts is extensive, precise, and extremely important for the taxonomist but hardly practical for anyone unused to the subject. It is well, however, to know that shrubs produce three main types of flowers:

(1) Perfect: both male and female parts in the same flower

(2) Male (staminate)

(3) Female (pistillate)

Some shrubs bear only perfect flowers; others are *monoecious,* producing both male and female flowers on the same plant. Some are *dioecious,* bearing male flowers on one plant and female ones on a separate plant. The dioecious shrubs, of course, bear fruit only on the female plants; the other two types can produce fruit on all fertile plants. There are occasional plants which have combinations of the above possibilities, but usually these require two or more plants to produce fruit and so for practical purposes are essentially dioecious. It is also true that, regardless of the type of flowers, fertilization is best achieved, and therefore plants are more productive of fruit, when several plants of the same species are present. Although the blooming period of most plants is brief, all mature shrubs should produce flowers, whereas not all bear fruit.

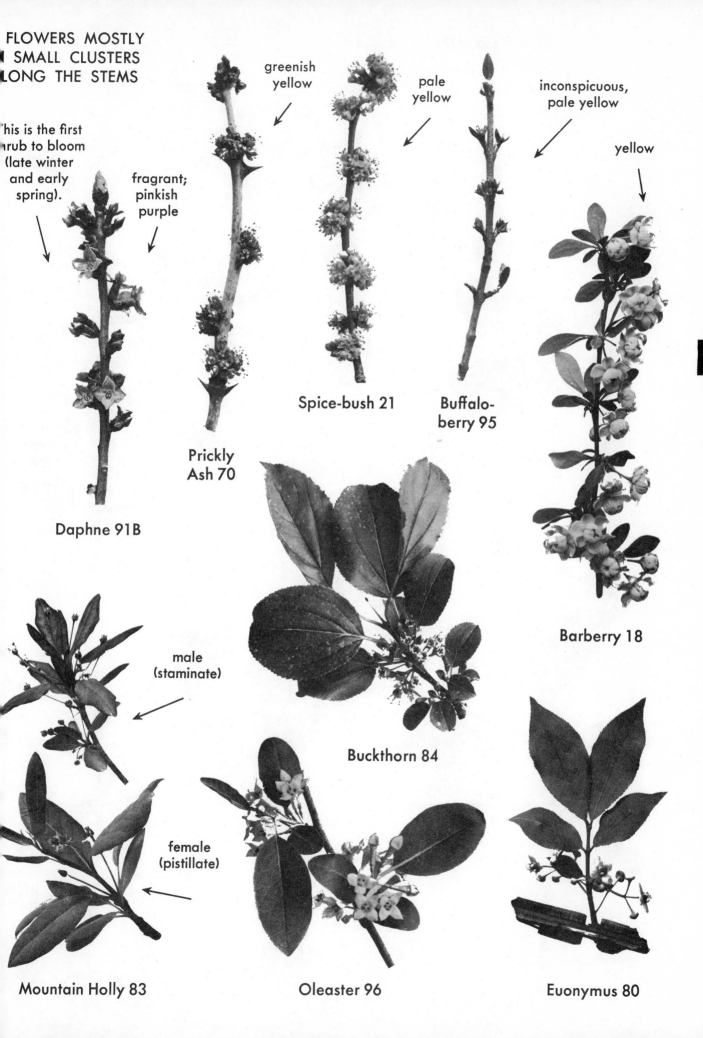

**FLOWERS MOSTLY
IN SMALL CLUSTERS
ALONG THE STEMS**

This is the first shrub to bloom (late winter and early spring).

fragrant; pinkish purple

greenish yellow

pale yellow

inconspicuous, pale yellow

yellow

Spice-bush 21

Buffalo-berry 95

Prickly Ash 70

Daphne 91B

Barberry 18

male (staminate)

Buckthorn 84

female (pistillate)

Mountain Holly 83

Oleaster 96

Euonymus 80

FL 1

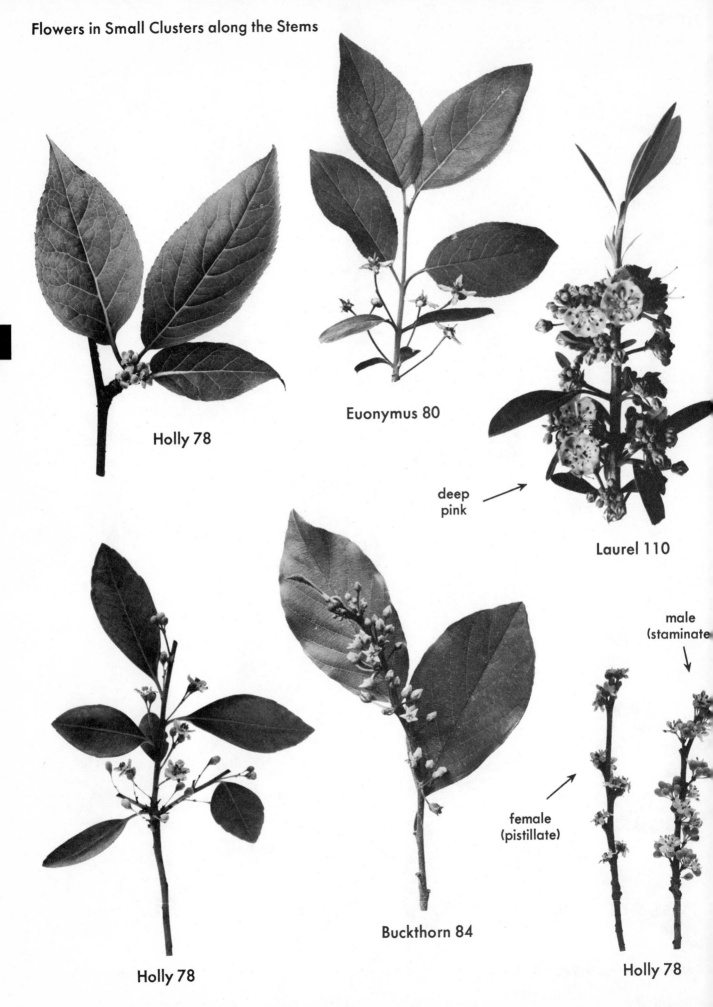

FL
2

Holly 78

Euonymus 80

deep
pink

Laurel 110

Holly 78

Buckthorn 84

male
(staminate

female
(pistillate)

Holly 78

Flowers in Small Clusters along the Stems

Both male and female flowers are produced in each individual cluster.

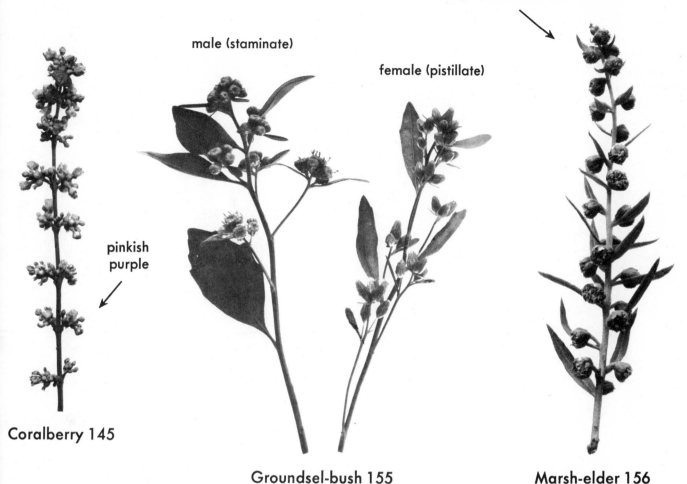

male (staminate)

female (pistillate)

pinkish purple

Coralberry 145

Groundsel-bush 155

Marsh-elder 156

FL 3

yellow

These are the last two shrubs to bloom (late in the fall).

rose purple

Witch-Hazel 38

Swamp Loose-strife 97B

Mistletoe 17A

II. FLOWERS IN THE FORM OF CATKINS

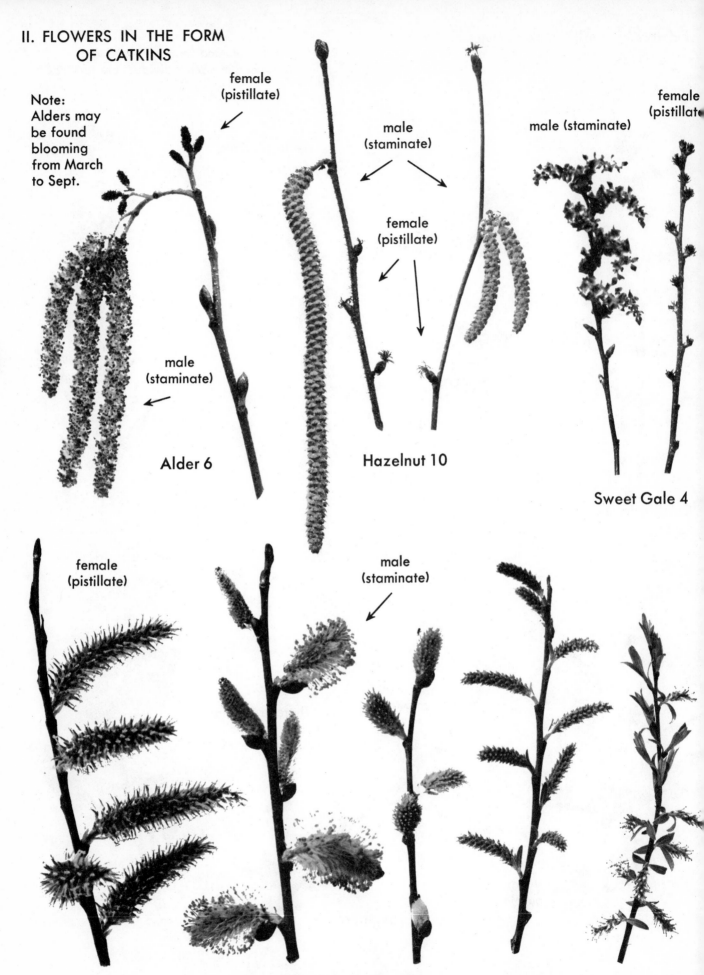

Note:
Alders may be found blooming from March to Sept.

female (pistillate)

male (staminate)

Alder 6

male (staminate)

female (pistillate)

Hazelnut 10

male (staminate)

female (pistillate)

Sweet Gale 4

female (pistillate)

male (staminate)

This row is all Willow 1.

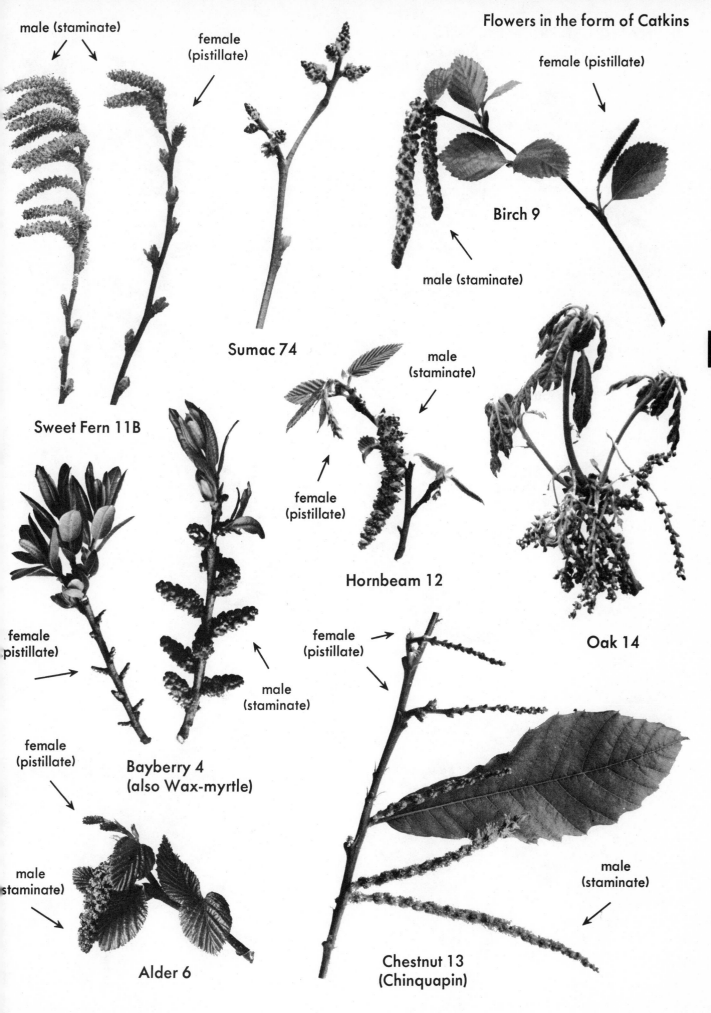

male (staminate)

female (pistillate)

female (pistillate)

Birch 9

male (staminate)

Sumac 74

FL 5

male (staminate)

female (pistillate)

Sweet Fern 11B

Hornbeam 12

Oak 14

female pistillate)

male (staminate)

female (pistillate)

female (pistillate)

male staminate)

Bayberry 4 (also Wax-myrtle)

male (staminate)

Alder 6

Chestnut 13 (Chinquapin)

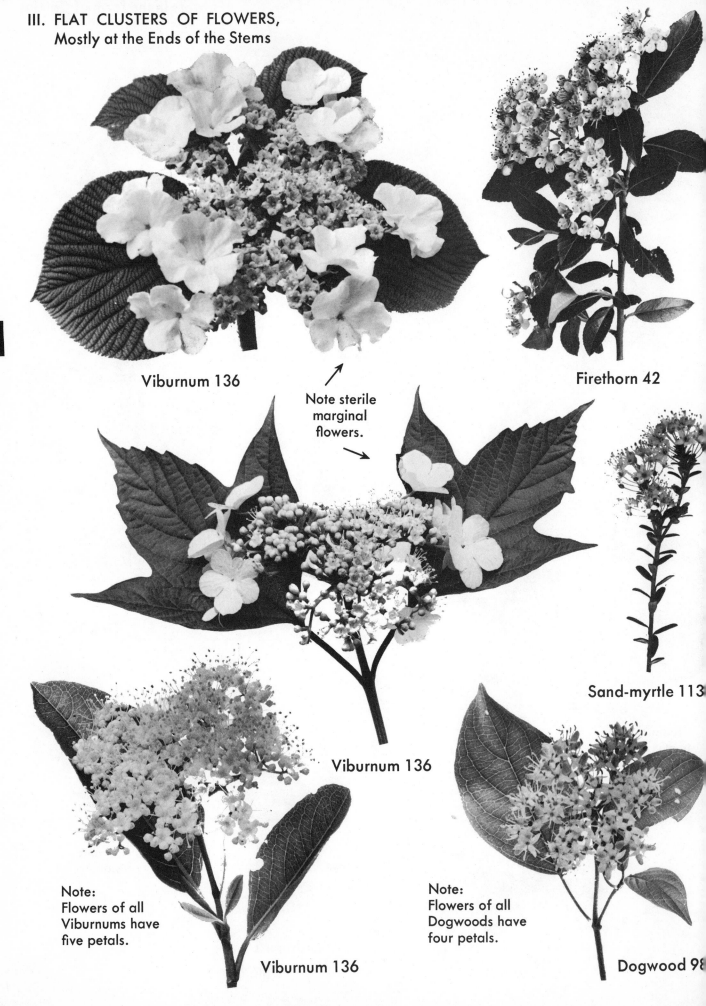

III. FLAT CLUSTERS OF FLOWERS,
Mostly at the Ends of the Stems

FL 6

Viburnum 136

Firethorn 42

Note sterile marginal flowers.

Viburnum 136

Sand-myrtle 113

Note:
Flowers of all Viburnums have five petals.

Note:
Flowers of all Dogwoods have four petals.

Viburnum 136

Dogwood 98

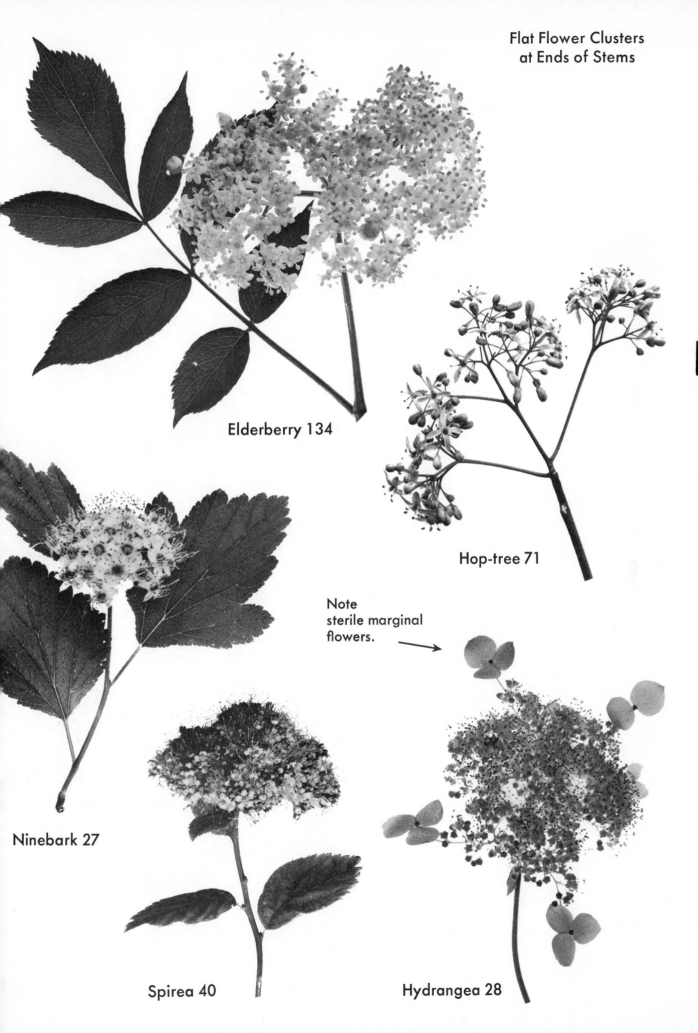

Elderberry 134

Hop-tree 71

FL
7

Note
sterile marginal
flowers.

Ninebark 27

Spirea 40

Hydrangea 28

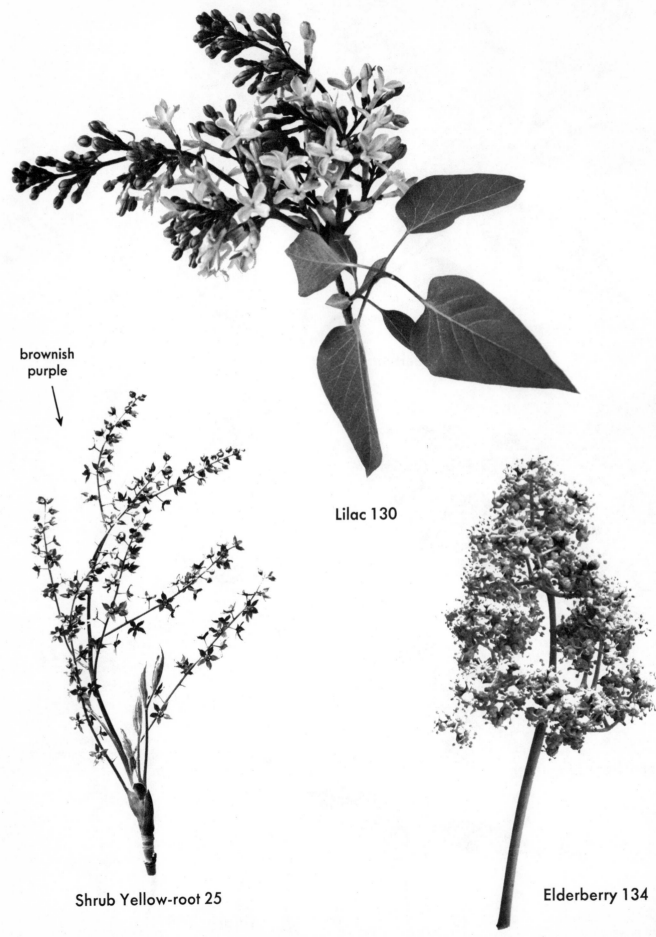

FL
8

brownish
purple

Lilac 130

Shrub Yellow-root 25

Elderberry 134

Rounded or Elongated Flower Clusters at Ends of Stems

Cherry 54

Currant 30

Currant 30

yellow

Barberry 18

Buffalo-nut 16

Note tiny pink or brownish hairs.

Fringe-tree 129

Smoke-tree 72

FL
9

purple with
golden pollen

FL
10

Virginia-willow 23B

False
Indigo 66

Dogwood 98

Spirea 40

New Jersey
Tea 45A

Privet 132

Deutzia 26

Rounded or Elongated Flower Clusters at Ends of Stems

Beware of this shrub; it is worse than Poison Ivy for many people.

Buttonbush 44

Privet 132

Poison Sumac 74

buds covered with sticky red hairs

pink

False Spirea 24

Bramble 46

Spirea 40

Clethra 102

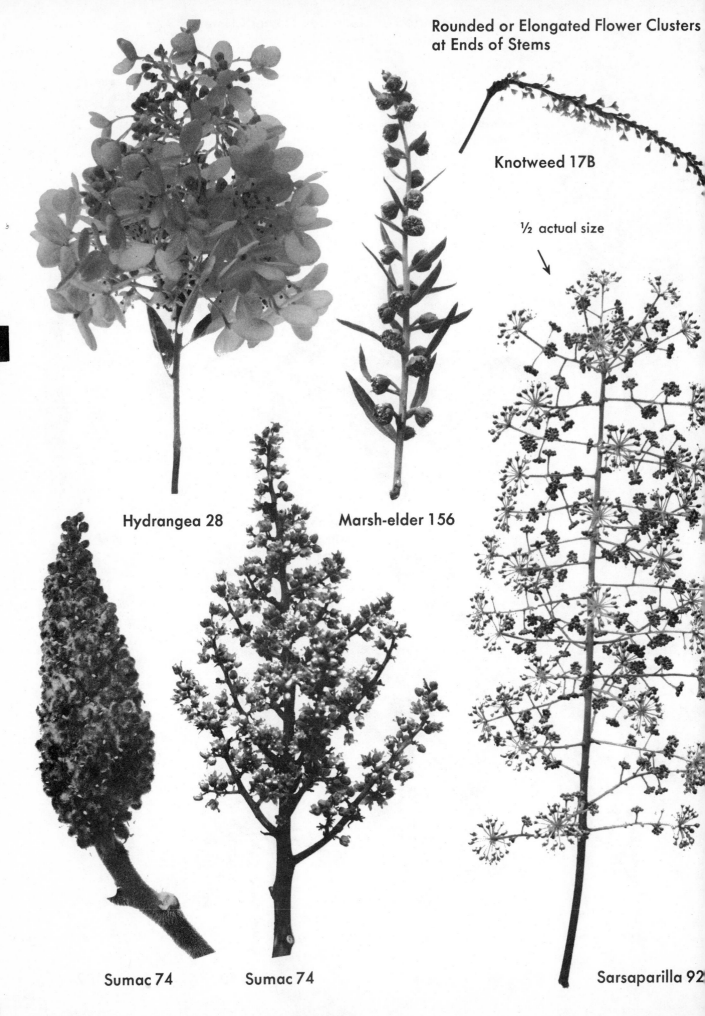

Rounded or Elongated Flower Clusters
at Ends of Stems

Knotweed 17B

½ actual size

Hydrangea 28

Marsh-elder 156

Sumac 74

Sumac 74

Sarsaparilla 92

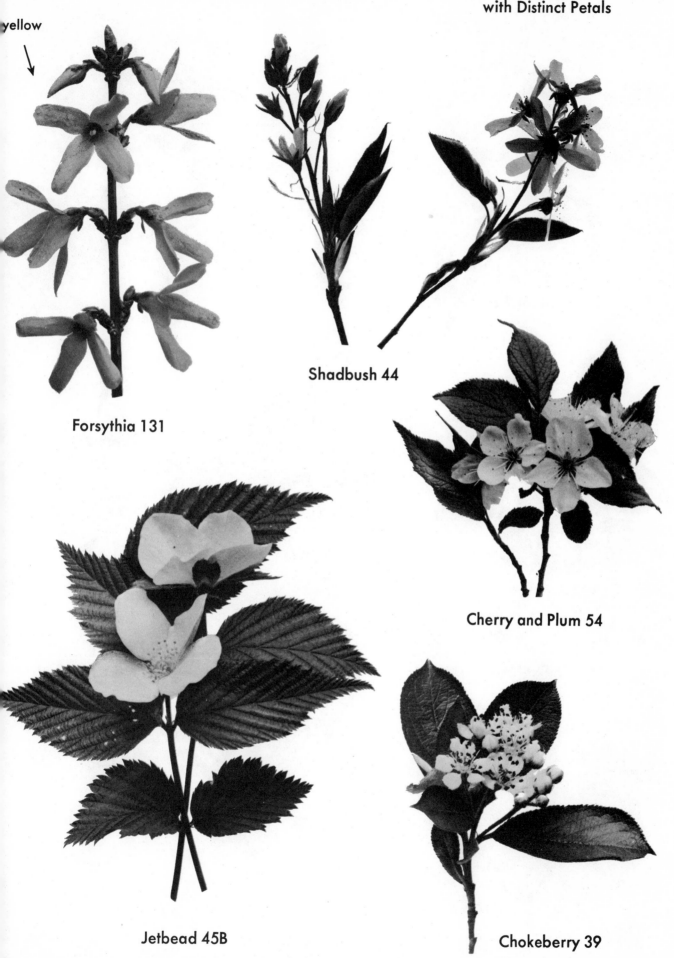

yellow

Forsythia 131

Shadbush 44

FL
13

Cherry and Plum 54

Jetbead 45B

Chokeberry 39

Widely Open Flowers

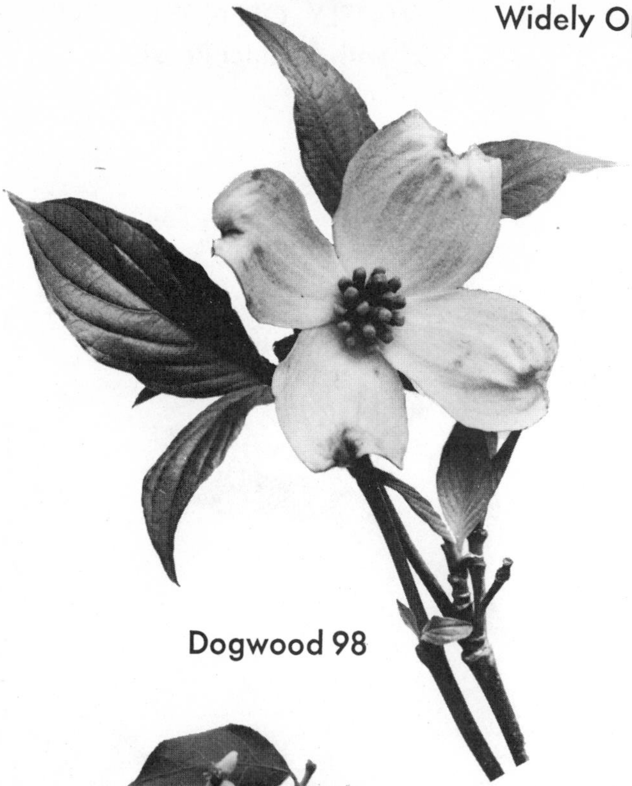

Dogwood 98

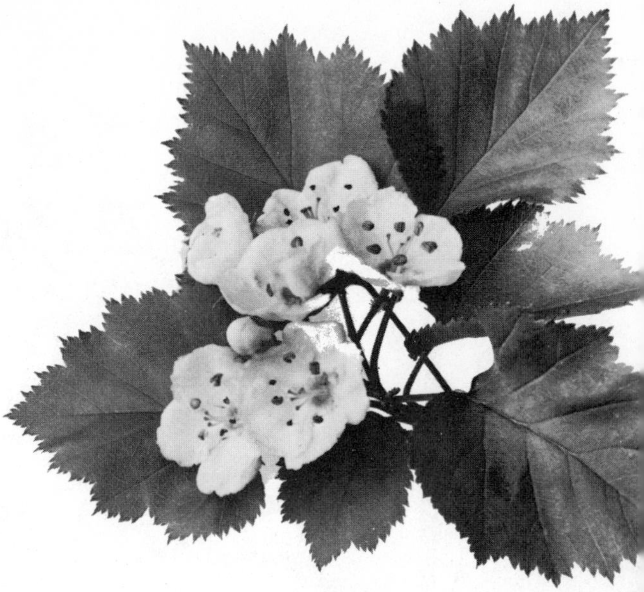

Hawthorn 43

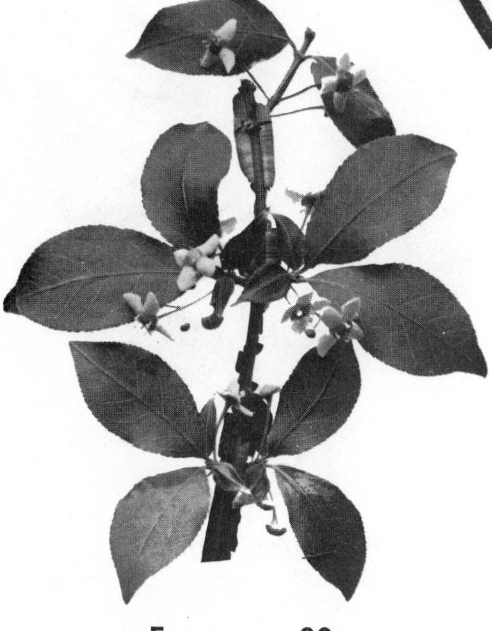

Euonymus 80

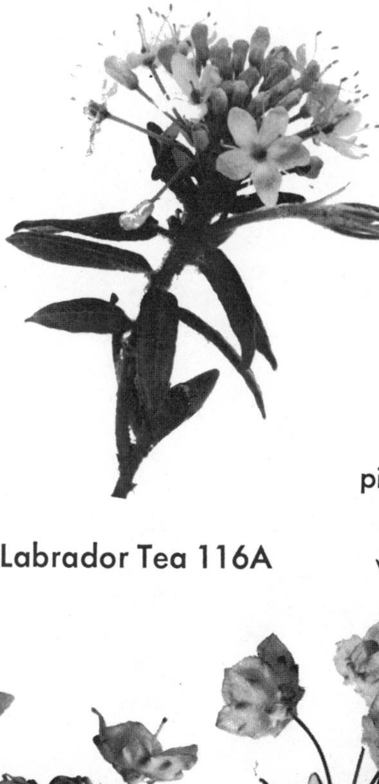

Labrador Tea 116A

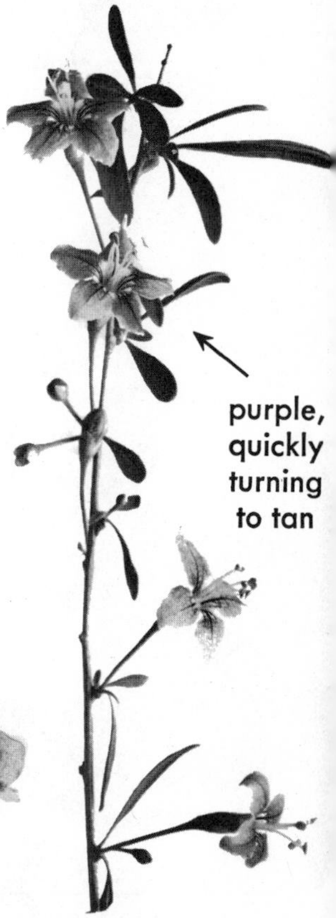

purple, quickly turning to tan

Matrimony-vine 125B

pink

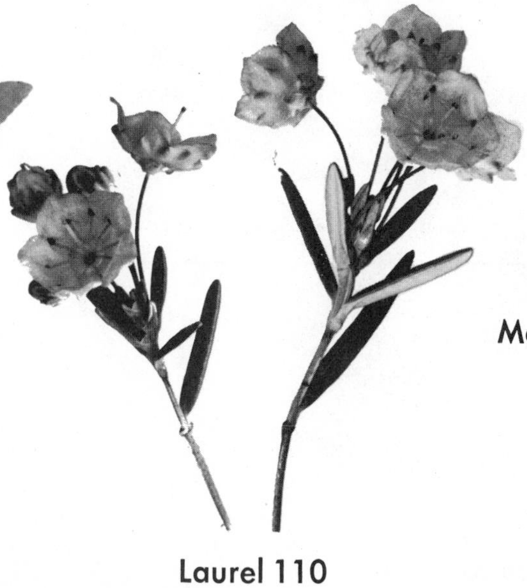

Laurel 110

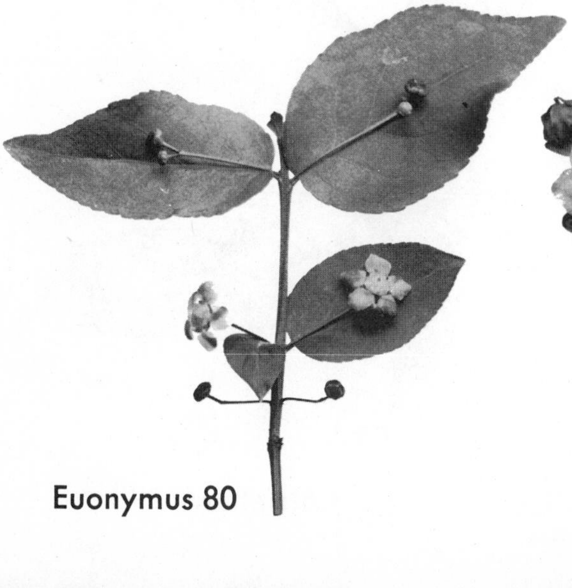

Euonymus 80

Bramble 46
(Raspberry)

FL
15

Bramble 46
(Blackberry)

Laurel 110

Mock-orange 22

Euonymus 80

Widely Open Flowers

magenta

Most wild
roses are
pink or white.

Rose 57

Rose 57

Rose 57

buds and flower
stems covered
with sticky
red hairs

magenta, fading
to lavender pink

**Bramble 46
(Raspberry)**

**Bramble 46
(Raspberry)**

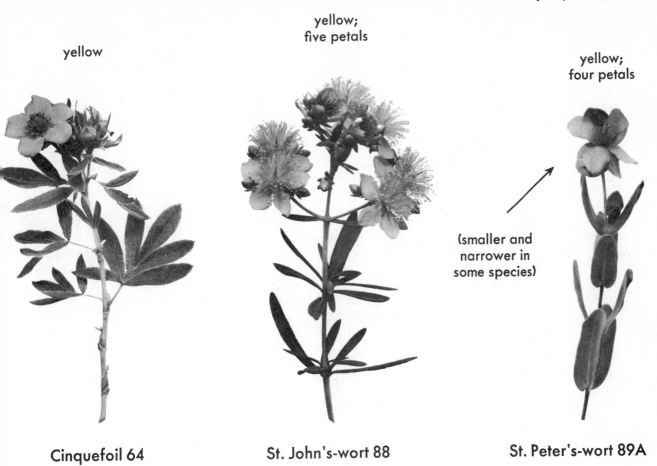

yellow

yellow;
five petals

yellow;
four petals

(smaller and
narrower in
some species)

FL
17

Cinquefoil 64

St. John's-wort 88

St. Peter's-wort 89A

double and single
forms

Color varies
from white to
pink, rose, blue
or purple.

Rose of Sharon 90

FL 18

Leatherleaf 116B

Andromeda 114

yellow

Leatherwood 91A

Andromeda 114

Huckleberry 126

Silverbell 128

Bladdernut 87

Blueberry 119

Blueberry 119

Blueberry 119
(Deerberry)

Bell-Shaped Flowers

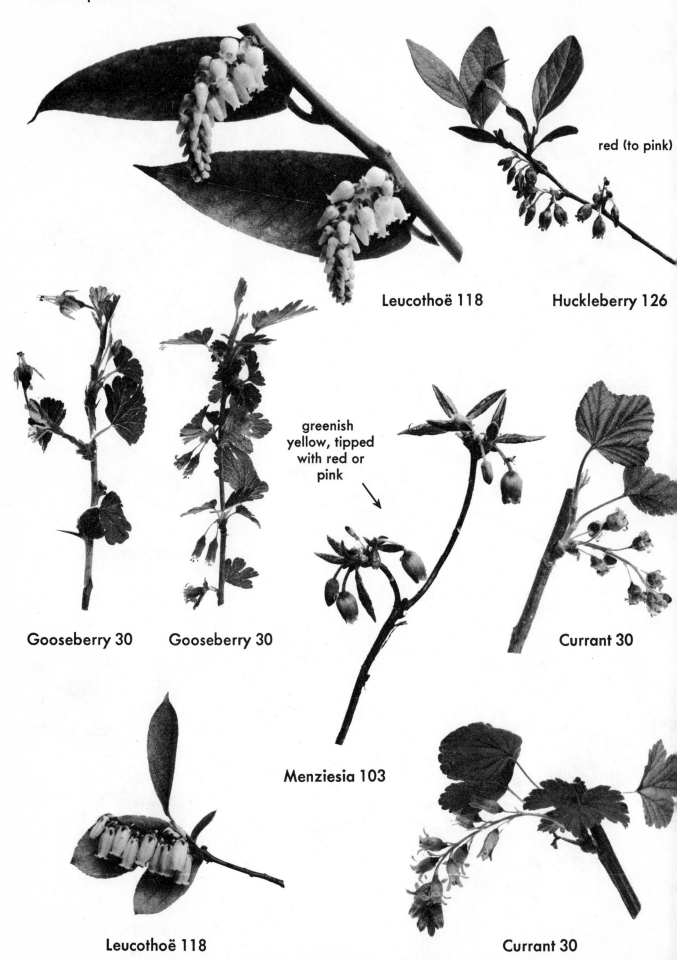

red (to pink)

Leucothoë 118

Huckleberry 126

Gooseberry 30

Gooseberry 30

greenish
yellow, tipped
with red or
pink

Currant 30

Menziesia 103

Leucothoë 118

Currant 30

Bog Rosemary 113A

Huckleberry 126

Huckleberry 126

Lyonia 117

pink

pinkish
purple

Snowberry 145

Blueberry 119
(Bilberry)

Lyonia 117

Coralberry 145

VII. TRUMPET-SHAPED OR DEEPLY CUT FLOWERS

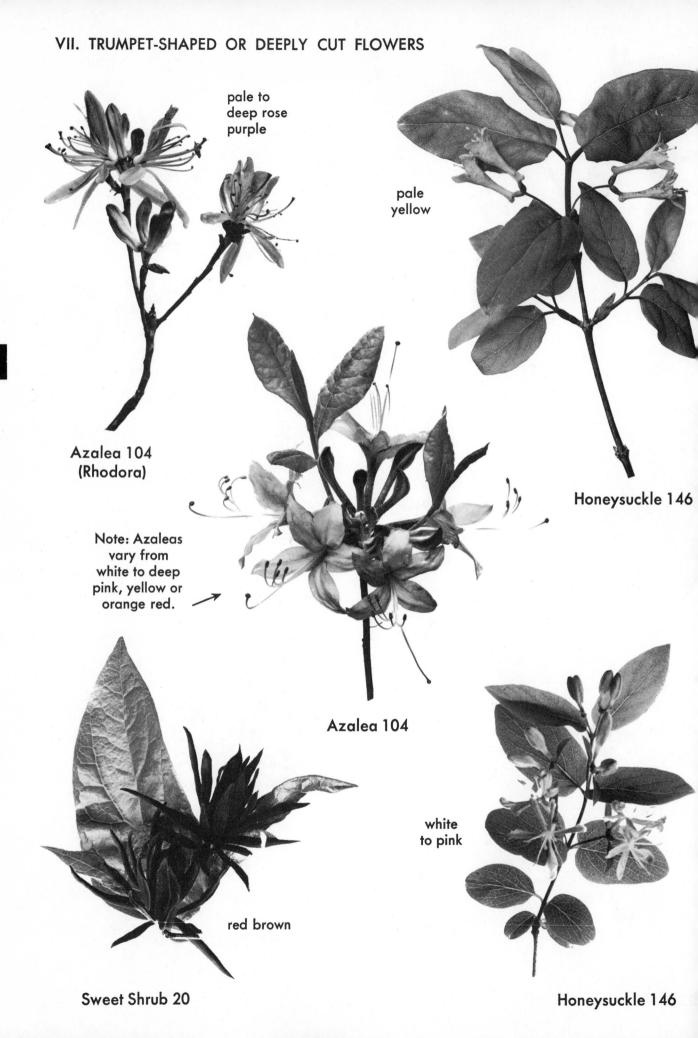

pale to
deep rose
purple

pale
yellow

**Azalea 104
(Rhodora)**

Honeysuckle 146

Note: Azaleas
vary from
white to deep
pink, yellow or
orange red.

Azalea 104

white
to pink

red brown

Sweet Shrub 20

Honeysuckle 146

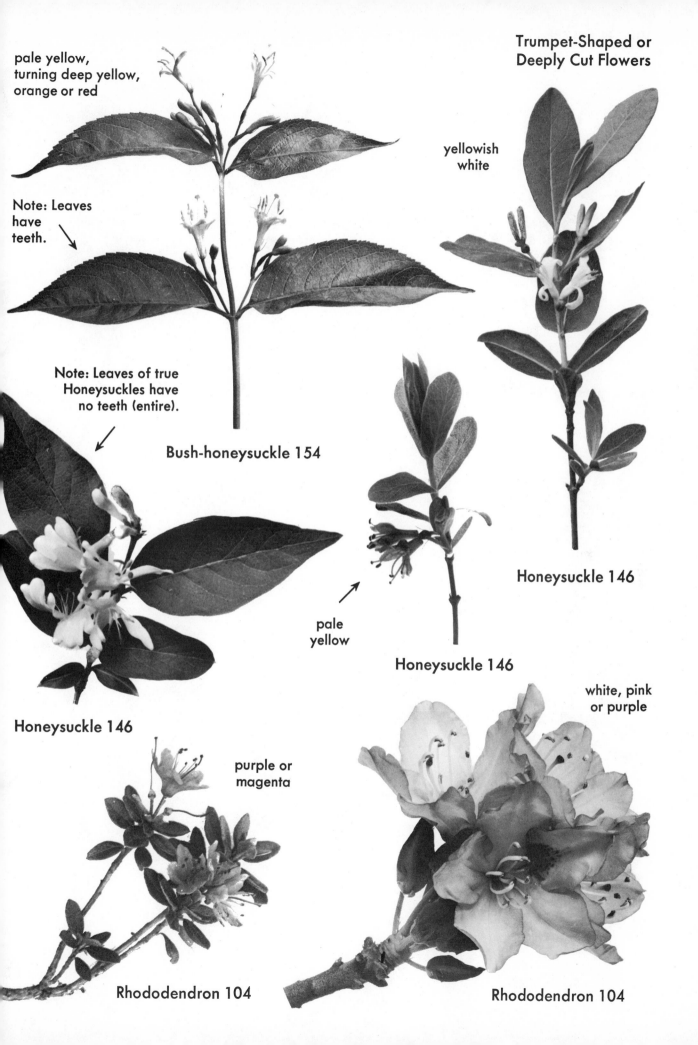

pale yellow,
turning deep yellow,
orange or red

yellowish
white

Note: Leaves
have
teeth.

Note: Leaves of true
Honeysuckles have
no teeth (entire).

FL
23

Bush-honeysuckle 154

Honeysuckle 146

Honeysuckle 146

pale
yellow

Honeysuckle 146

Honeysuckle 146

white, pink
or purple

purple or
magenta

Rhododendron 104

Rhododendron 104

VIII. PEA-TYPE FLOWERS

yellow

Gorse 67B

yellow

yellow

Scotch Broom 65B

reddish pink
to pale pink

Redbud 68

pink

Bristly Locust 69

yellow

Dyer's Greenweed 65

Key #4—FRUIT

This Key is arranged as follows:

Note: The first and fourth groups are short; the second and third are longer. A little study, however, will indicate that the individual fruits in both the capsule and berry-like sections are distinctive, and few, if any, are so alike as to cause confusion. Note in both these sections the type of growth of the whole cluster, whether branching from one point or from several points, or not branching and arranged along a single axis. Note shape and size of the individual capsules and berries. The word "bloom" is often used to describe a whitish, powdery covering often found on berries (and other parts of a plant). This bloom, when present, can be rubbed off revealing the true color beneath.

The fruit of many plants persists for a considerable part of the year and therefore is a valuable aid to identification; even a dried berry may be found in winter, which at least eliminates other types of fruit for that particular plant.

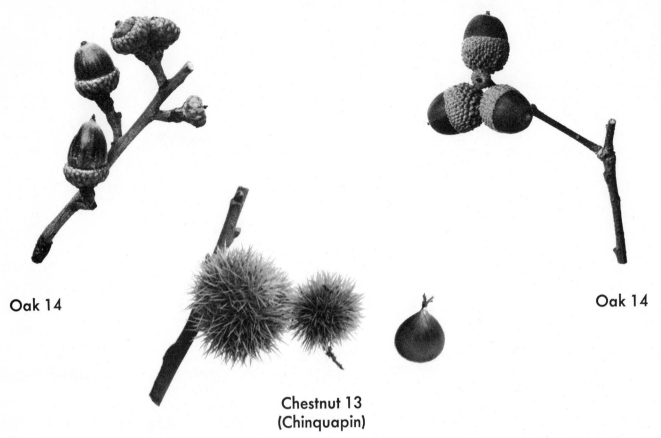

Oak 14

Oak 14

Chestnut 13
(Chinquapin)

poisonous

Buffalo-nut 16

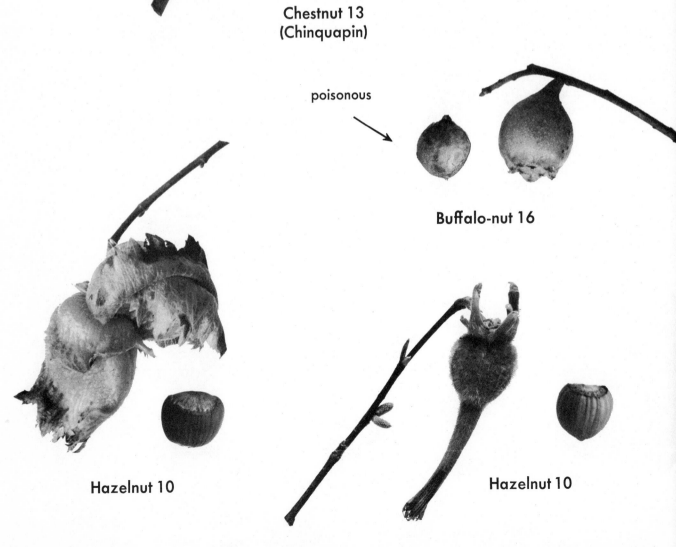

Hazelnut 10

Hazelnut 10

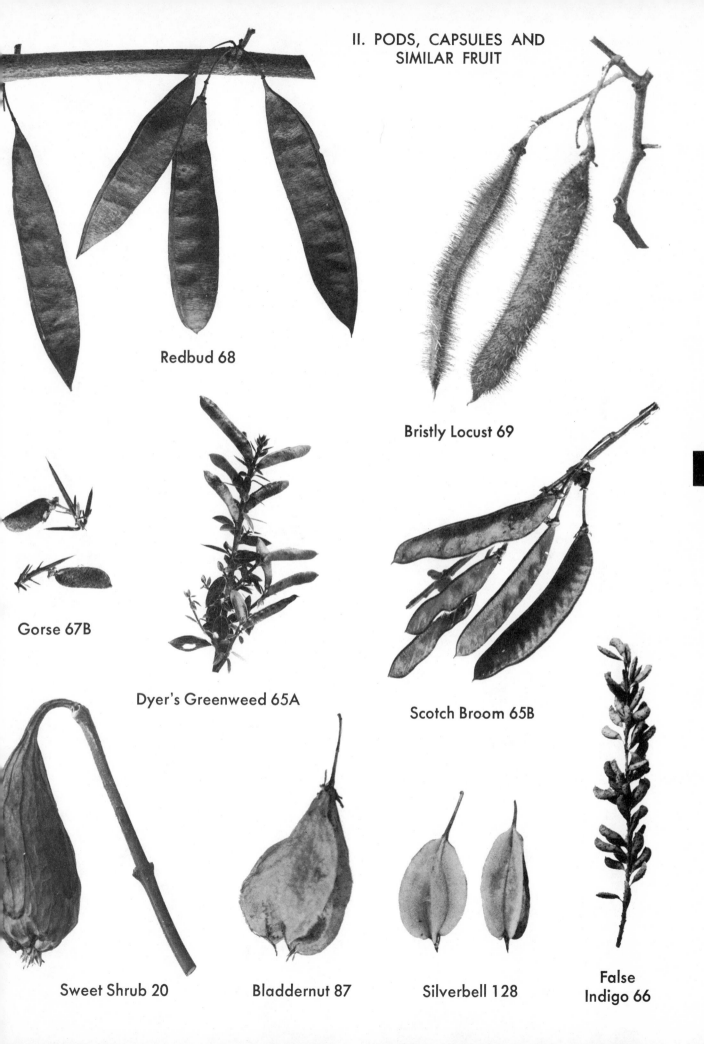

Redbud 68

Bristly Locust 69

Gorse 67B

Dyer's Greenweed 65A

Scotch Broom 65B

Sweet Shrub 20

Bladdernut 87

Silverbell 128

False
Indigo 66

FR
2

Pods, Capsules and Similar Fruit

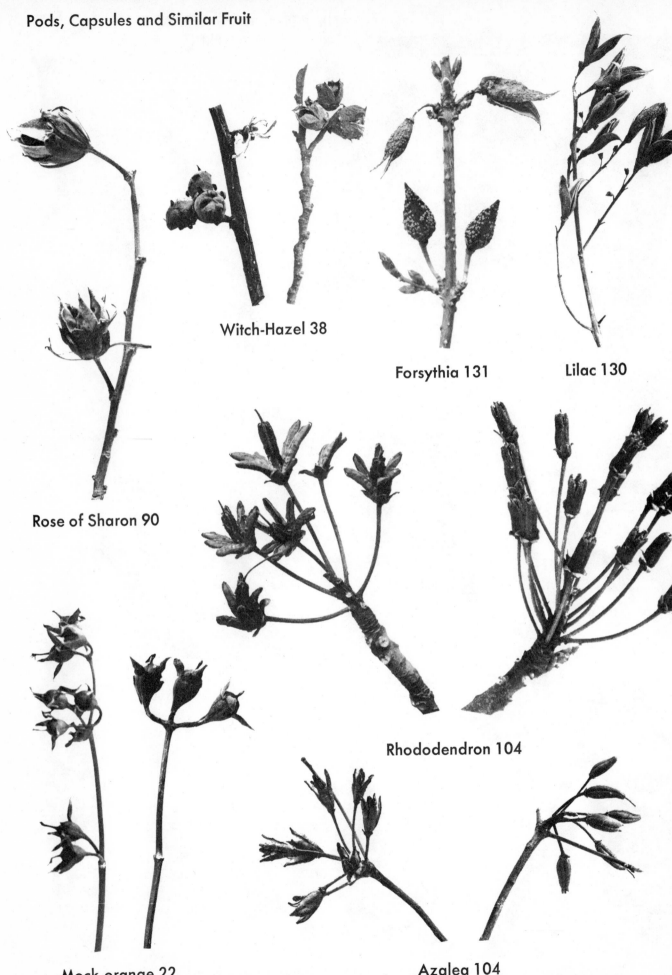

Witch-Hazel 38

Forsythia 131

Lilac 130

Rose of Sharon 90

Rhododendron 104

Mock-orange 22

Azalea 104

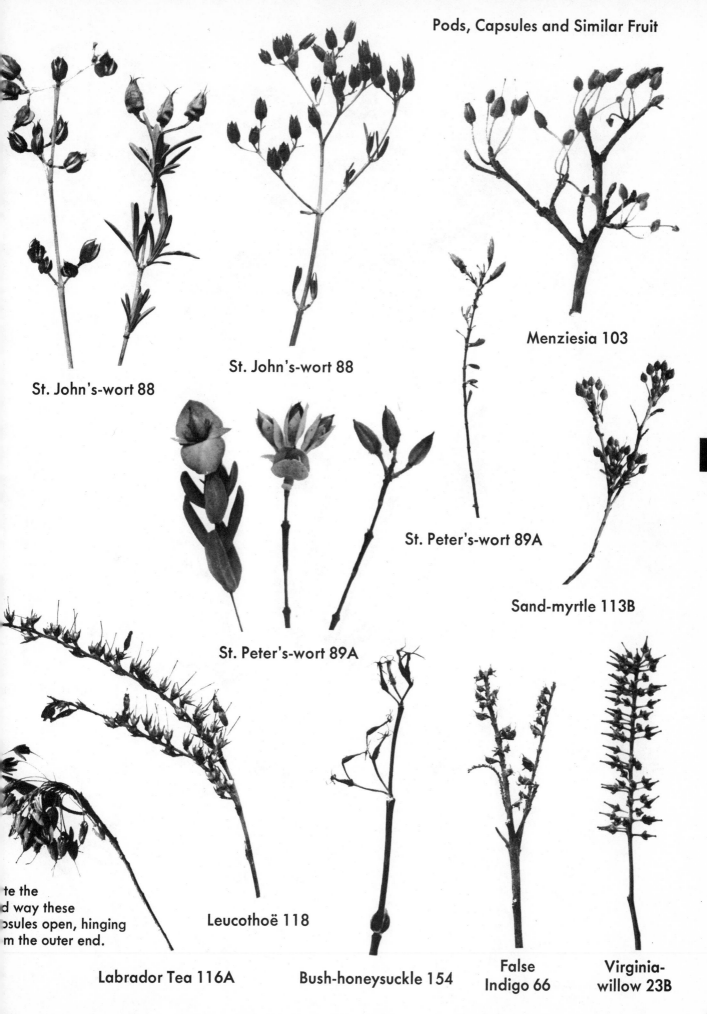

St. John's-wort 88

St. John's-wort 88

Menziesia 103

St. Peter's-wort 89A

St. Peter's-wort 89A

Sand-myrtle 113B

FR
4

te the
d way these
osules open, hinging
m the outer end.

Leucothoë 118

Labrador Tea 116A

Bush-honeysuckle 154

False
Indigo 66

Virginia-
willow 23B

Laurel 110

Laurel 110

FR
5

Andromeda 114

Leatherleaf 116

Laurel 110

Andromeda 114

Bog Rosemary 113A

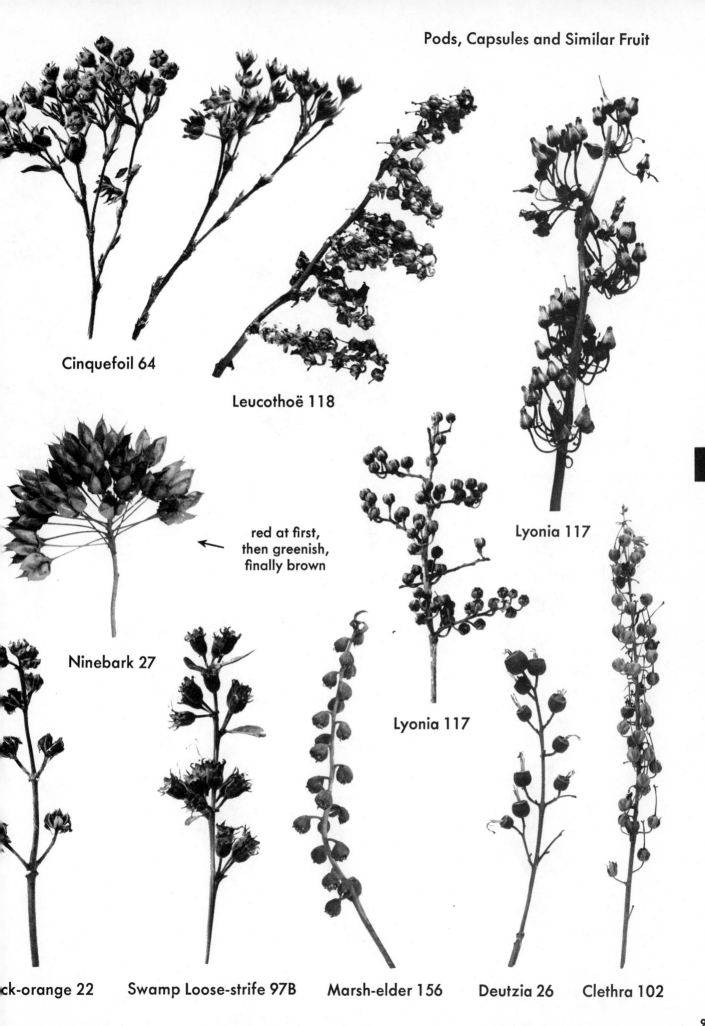

Cinquefoil 64

Leucothoë 118

Ninebark 27

red at first,
then greenish,
finally brown

Lyonia 117

Lyonia 117

FR
6

ck-orange 22   Swamp Loose-strife 97B   Marsh-elder 156   Deutzia 26   Clethra 102

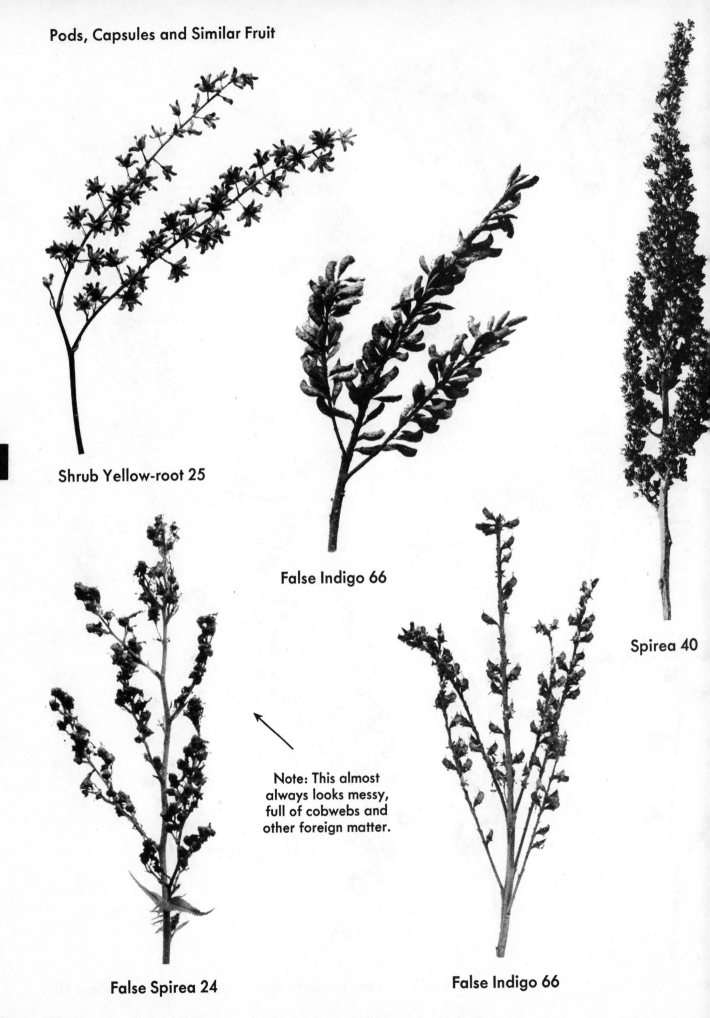

FR
7

Shrub Yellow-root 25

False Indigo 66

Spirea 40

Note: This almost
always looks messy,
full of cobwebs and
other foreign matter.

False Spirea 24

False Indigo 66

remnants of
sterile flowers

few actual
fruit

Spirea 40

Hydrangea 28

Spirea 40

Hydrangea 28

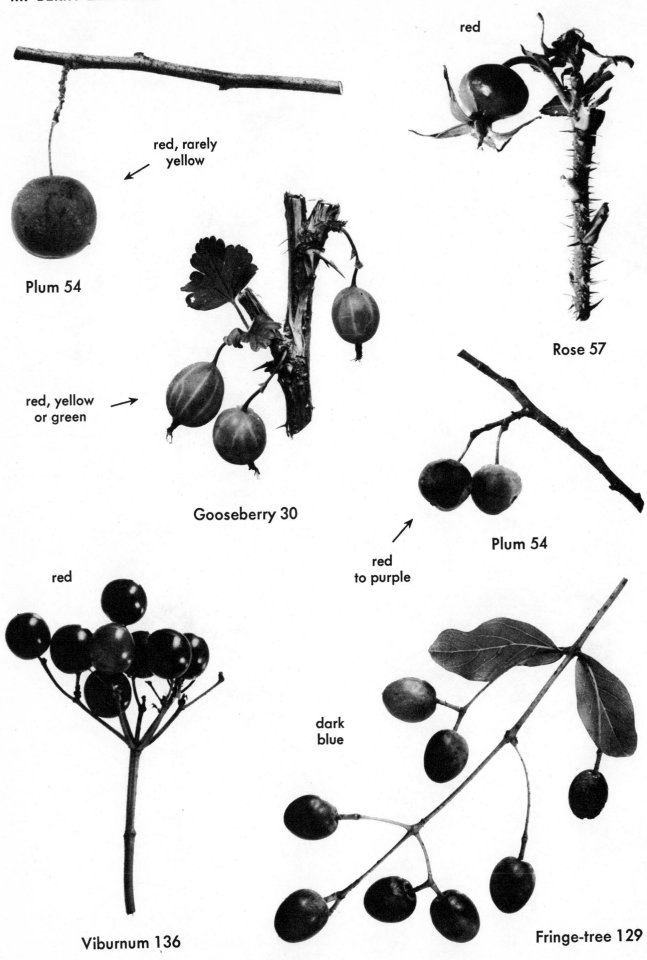

red

Rose 57

red, rarely
yellow

Plum 54

red, yellow
or green

Gooseberry 30

Plum 54

red
to purple

red

Viburnum 136

dark
blue

Fringe-tree 129

FR
9

yellow to pink,
finally dark
blue with bloom

red

blue

Viburnum 136

Dogwood 98

Rose 57

orange red

white

red, turning
dark purple or black

Firethorn 42

Dogwood 98

Viburnum 136

Berry-like Fruit

black

Elderberry 134

dark
blue

Viburnum 136

three types:
red, purple
or black

←

Chokeberry 39

dark blue
or black,
sometime
with bloo

Viburnum 136

shiny
black

Viburnum 136

pink, turning
dark blue wit
heavy bloom

Viburnum 136

red

shiny
black

black with bloom,
giving bluish cast

Privet 132

Privet 132

Privet 132

FR
12

blue black
on red
stems

Elderberry 134

red

Dogwood 98

white

pink, turning
dark blue
with bloom

Dogwood 98

Rose 57

Viburnum 136

Berry-like Fruit

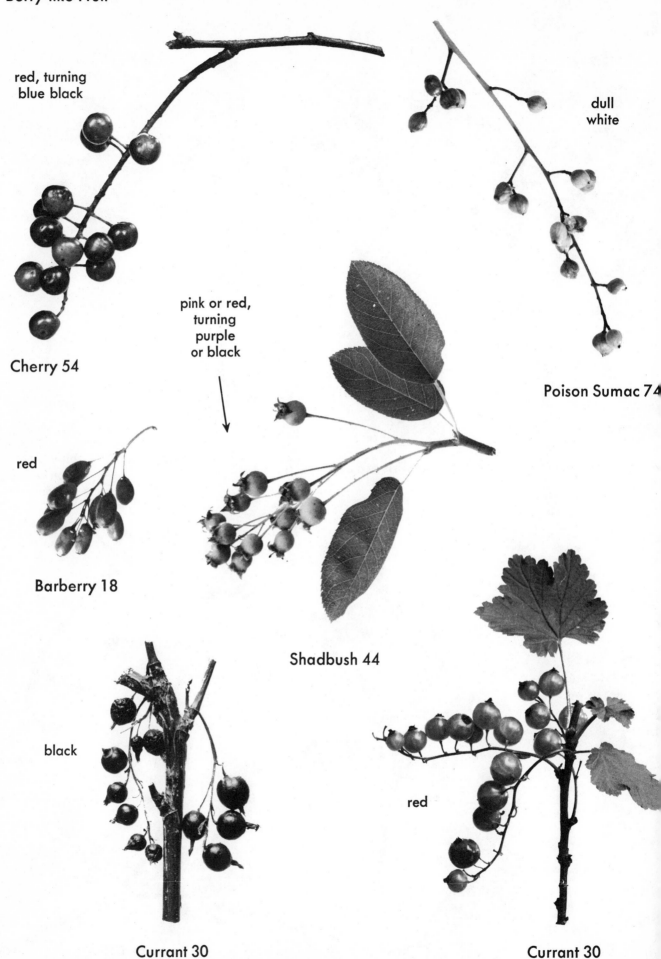

red, turning
blue black

Cherry 54

dull
white

Poison Sumac 74

pink or red,
turning
purple
or black

red

Barberry 18

Shadbush 44

black

red

Currant 30

Currant 30

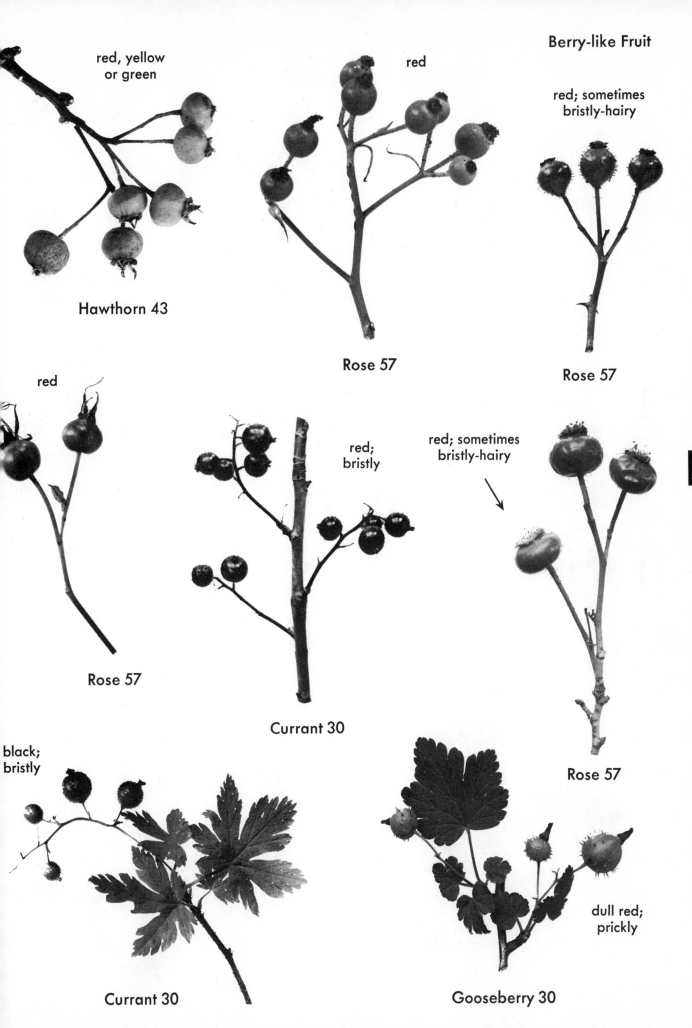

red, yellow
or green

red

red; sometimes
bristly-hairy

Hawthorn 43

Rose 57

Rose 57

red

red;
bristly

red; sometimes
bristly-hairy

FR
14

Rose 57

Currant 30

Rose 57

black;
bristly

dull red;
prickly

Currant 30

Gooseberry 30

105

Berry-like Fruit

red

red

red

Mountain Holly 83

Honeysuckle 146

Honeysuckle 146

All Blueberries
contain many
small seeds.

All Huckleberries
are ten-seeded.

blue with blo

greenish (or slightly
purplish) with bloom

Huckleberry 126

Blueberry 119
(Deerberry)

dull red

dull
red

blue
with
bloom

Honeysuckle 146

Gooseberry 30

Gooseberry 30

FR
15

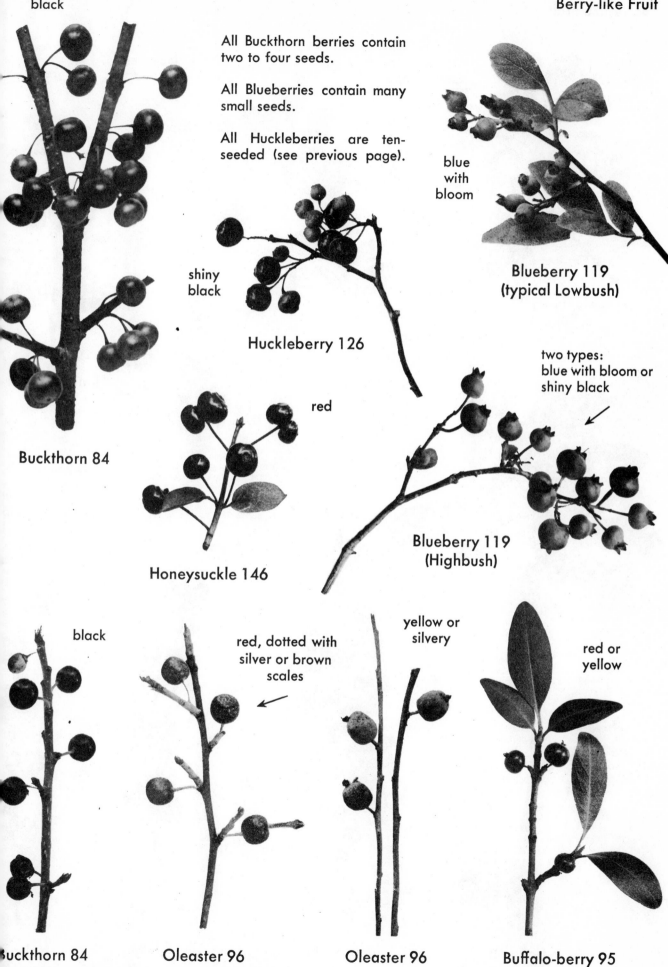

black

All Buckthorn berries contain two to four seeds.

All Blueberries contain many small seeds.

All Huckleberries are ten-seeded (see previous page).

blue with bloom

Blueberry 119 (typical Lowbush)

shiny black

Huckleberry 126

two types: blue with bloom or shiny black

Buckthorn 84

red

FR 16

Honeysuckle 146

Blueberry 119 (Highbush)

black

red, dotted with silver or brown scales

yellow or silvery

red or yellow

Buckthorn 84   Oleaster 96   Oleaster 96   Buffalo-berry 95

Berry-like Fruit

blue
with
bloom

All Blueberries
contain many
small seeds.

All Huckleberries
are ten-seeded.

red

blue
with
bloom

**Blueberry 119
(Bilberry)**

black

Huckleberry 126

Honeysuckle 1

FR
17

Huckleberry 126

black

Holly 78

red

red

red

Holly 78

Daphne 91B

Honeysuckle 146

Holly 7

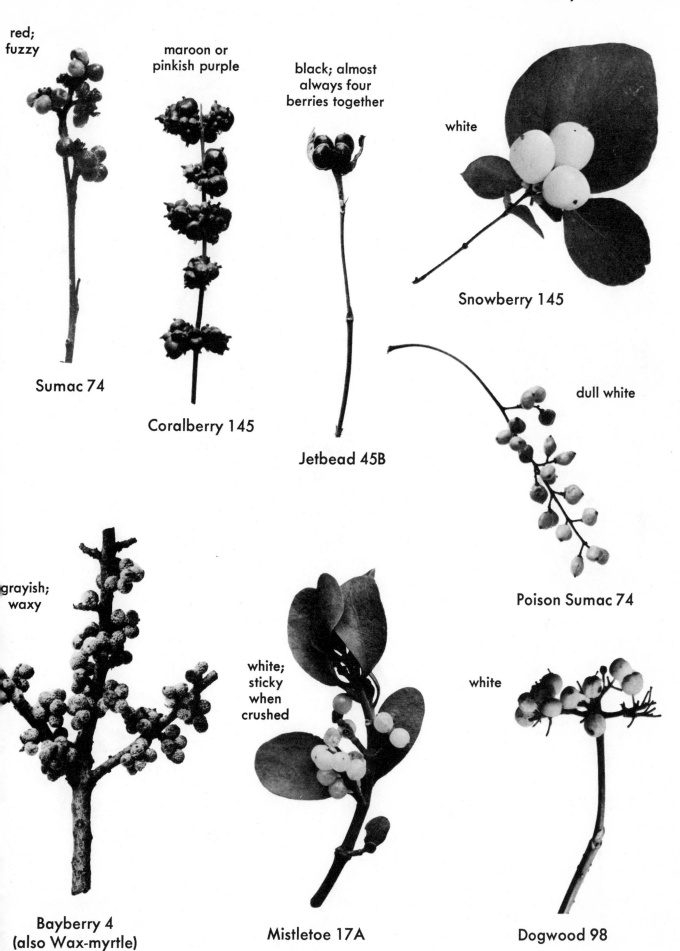

red;
fuzzy

maroon or
pinkish purple

black; almost
always four
berries together

white

Snowberry 145

Sumac 74

Coralberry 145

Jetbead 45B

dull white

FR
18

Poison Sumac 74

grayish;
waxy

white;
sticky
when
crushed

white

Bayberry 4
(also Wax-myrtle)

Mistletoe 17A

Dogwood 98

Berry-like Fruit

purple to
black

red

yellow to
red

red

Cherry 54

Matrimony-vine 125B

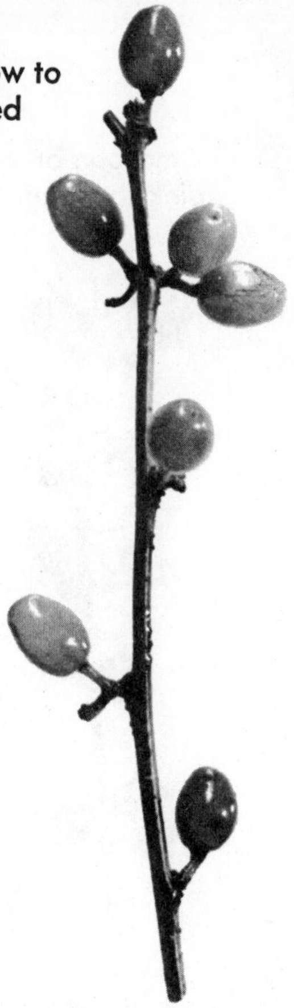

silvery

green, yellowish
or red

Barberry 18

Oleaster 96

Spice-bush 21

red

red

Leatherwood 91A

Barberry 18

red

black

Dogwood 98

red

blue
with
bloom

Honeysuckle 146

Euonymus 80

Viburnum 136

Honeysuckle 146

FR
19

Berry-like Fruit

seeds with orange skin

reddish pink

pink

nymus 80 (Euonymus 80)

black or red

Euonymus 80

black

red

black when ripe

ramble 46 (Raspberry)

Bramble 46 (Raspberry)

½ actual size

FR 20

Sarsaparilla 92

black

½ actual size

Bramble 46 (Blackberry)

red

Bramble 46 (Raspberry)

Sarsaparilla 92

IV. MISCELLANEOUS FRUIT

Seed heads produce cotton (pappus) which is carried by the wind, distributing the seed.

winged seed

Hornbeam 12

winged seed

Knotweed 17B

Groundsel-bush 155

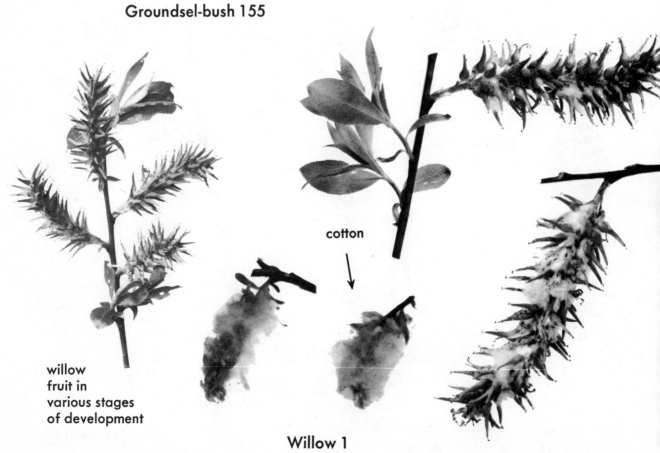

cotton

willow fruit in various stages of development

Willow 1

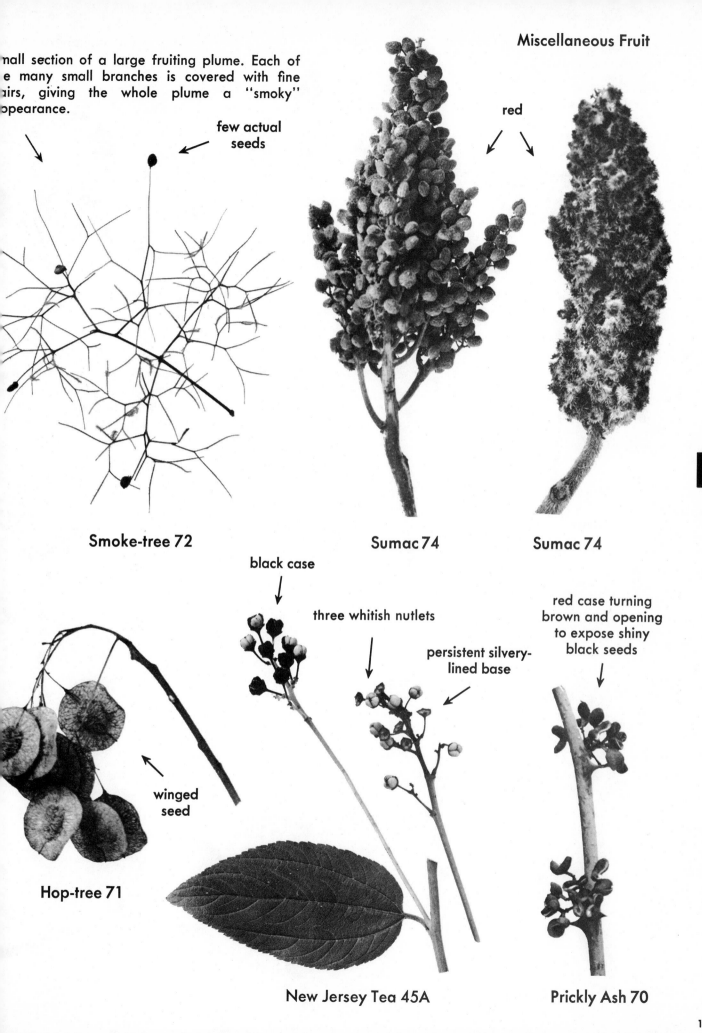

nall section of a large fruiting plume. Each of
e many small branches is covered with fine
airs, giving the whole plume a "smoky"
ppearance.

few actual
seeds

red

Smoke-tree 72

Sumac 74

Sumac 74

FR
22

black case

three whitish nutlets

persistent silvery-
lined base

red case turning
brown and opening
to expose shiny
black seeds

winged
seed

Hop-tree 71

New Jersey Tea 45A

Prickly Ash 70

113

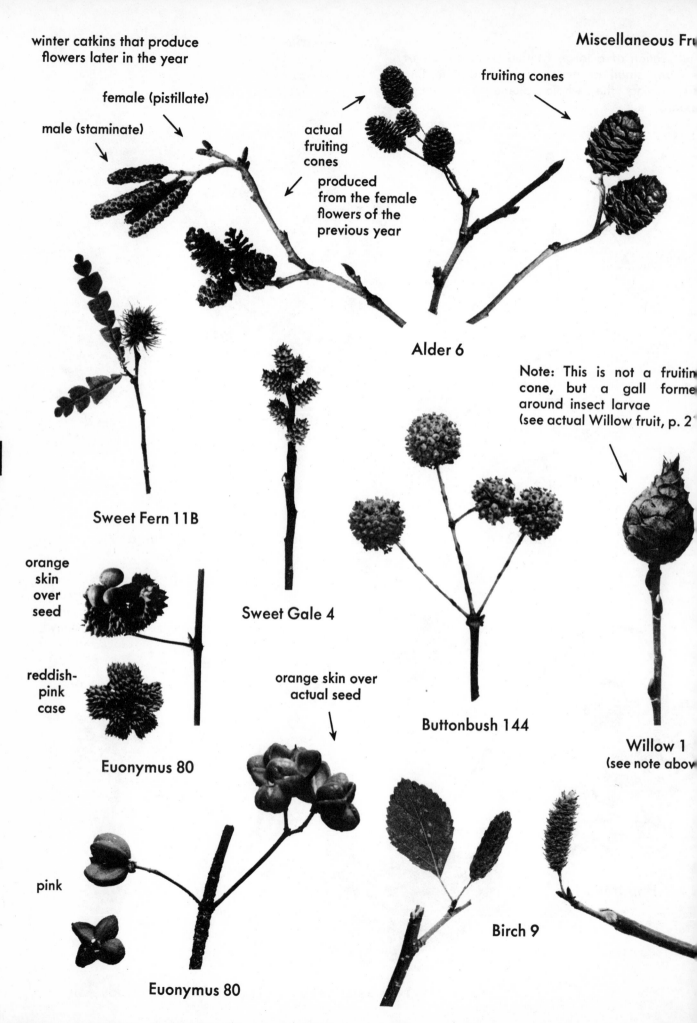

winter catkins that produce
flowers later in the year

female (pistillate)

male (staminate)

fruiting cones

actual
fruiting
cones

produced
from the female
flowers of the
previous year

Alder 6

Note: This is not a fruitin
cone, but a gall forme
around insect larvae
(see actual Willow fruit, p. 2

Sweet Fern 11B

orange
skin
over
seed

reddish-
pink
case

Sweet Gale 4

orange skin over
actual seed

Buttonbush 144

Willow 1
(see note abov

Euonymus 80

pink

Euonymus 80

Birch 9

FR
23

114

Key #5—TWIGS

(Evergreens not included)

This Key is divided into two main sections:

I ALTERNATE: buds (and leaf-scars) arranged alternately along the twigs, pp. 1-15

Note: Pages 1-4 include twigs which are remarkable for one reason or another. Beginning on page 5 they are arranged in order of size of twigs and buds, ending with twigs which die back in winter, or whose buds are inconspicuous or entirely invisible, being concealed within the twig itself.

II OPPOSITE: buds (and leaf-scars) opposite, or in whorls of three or four around the twig at each node (growth point), pp. 16-21

Leaf-scars are left at the point of separation when the leaves drop off in the fall. They vary greatly in size and shape and are often typical of a particular plant.

Buds develop during the summer at the ends of twigs and also along the twigs, usually just above the junction of the leaf-stalk and twig (in the axil thus formed), and therefore are found just above the leaf-scar on the winter twig. A few plants produce buds inside the leaf-stalk itself and thus appear in the middle of the leaf-scar.

Buds, to the untrained eye, may appear much alike, but a little scrutiny reveals that each plant has its own particular type of bud, and therefore buds are an extremely important aid to identification. A lens shows these details clearly, but even without one many differences are apparent. Buds contain the whole potential of growth, leaves and flowers, and it is therefore no wonder that they are distinctive.

Buds from which flowers will emerge are often markedly different (usually larger) than leaf (or growth) buds, and when present are very characteristic of specific plants.

Buds at the ends of the twigs are often larger and more distinctive than the side (axillary) buds.

Buds are usually protected in winter by an outer layer of overlapping scales, the number, shape, arrangement and color of which are often characteristic. A few plants have "naked" buds (without scales), and among the shrubs included in this book the Willows are the only plants with buds covered by only a single scale.

Twigs (and buds) vary considerably in color, but for our purpose, unless note is made of color, it can be assumed that the twigs are some type of brown or brownish gray. Throughout this Key note is often made of red (maroon) *and* green twigs. The reason for this apparently startling difference of color on the same twig results from the fact that some twigs which are normally green in the shade turn red in the sun (or vice versa), and frequently one half of a twig will thus be red, the other (the side not exposed to the sun) will be green. As with other parts of a plant, twigs are often covered with a whitish bloom which can be rubbed off, revealing the true color beneath.

The word twig, as used here, applies largely to the latest growth of the shrub. This is easily determined, as the new growth of almost all shrubs is different in color and texture from that of the older wood, and the junction between the old and the new growth is almost always marked by a ring around the twig at that point. It should also be noted that almost all shrubs produce side branches only from two-year wood (or older). Two remarkable exceptions to this are seen on page 2. Both Alternate-leaved Dogwood and Clethra branch the first year and this in itself marks these two plants apart from other shrubs.

I. ALTERNATE SECTION

long, pointed buds; side buds
close to stem (appressed)

brown buds;
twigs smooth

red buds;
twigs smooth

red buds;
twigs fuzzy

The buds of all Willows h
only one scale. (They are
only plants in this book w
this characteristic.)

Shadbush 44   Chokeberry 39   Chokeberry 39   Willow 1

These two plants have naked buds (buds without any
scales). (They are the only shrubs shown here with alternate
buds that have this characteristic.)

dark brown buds m
opposite, but occasion
partly alternate (see
Fringe-tree, p. 5)

Witch-Hazel 38   Buckthorn 84   Buckthorn 84

TW
1

Alternate Section

These two plants produce side branches on the new growth.
(They are the only plants in this book that normally do this.)

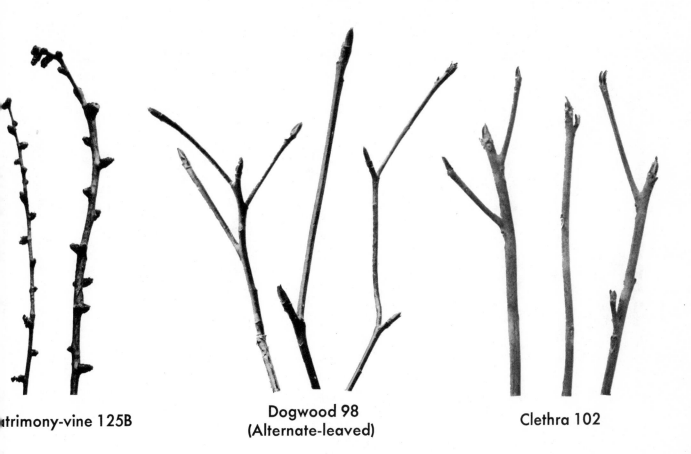

trimony-vine 125B

Dogwood 98
(Alternate-leaved)

Clethra 102

These two plants usually have spurs on which flowers
and fruit are produced.

Bayberry 4

Holly 78

Mountain Holly 83

Note: The plants on this page commonly pro‑
duce buds in clusters at the ends of the twigs

flower
buds

leaf
buds

buds covered with a
bloom, giving grayish
or pinkish look; scales
indistinct

hairy twigs

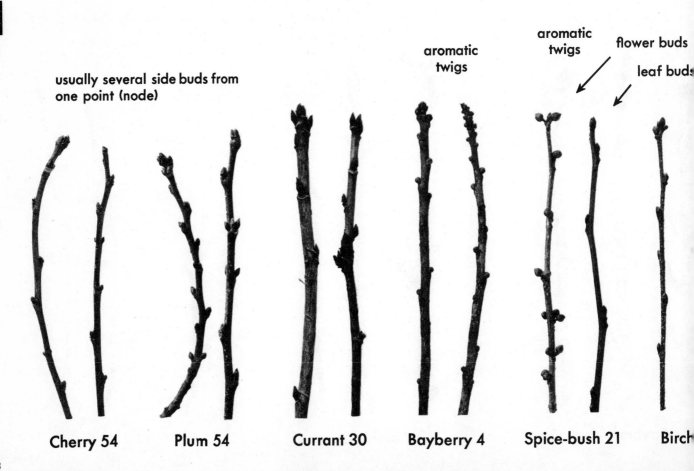

Azalea 104

**Azalea 104
(Rhodora)**

Menziesia 103

Oak 1

TW
3

aromatic
twigs

aromatic
twigs

flower buds

leaf buds

usually several side buds from
one point (node)

Cherry 54

Plum 54

Currant 30

Bayberry 4

Spice-bush 21

Birch

Note: The plants on this page commonly produce catkins, which are evident in winter. These develop later in the season as male (staminate) flowers. Some Alders also have catkins that develop into female (pistillate) flowers.

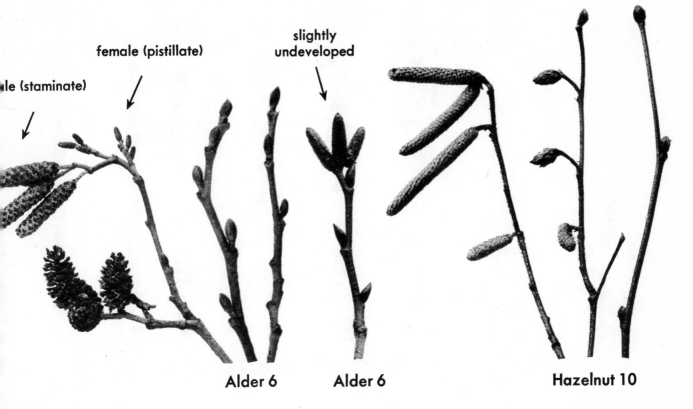

female (pistillate)

slightly
undeveloped

male (staminate)

Alder 6   Alder 6   Hazelnut 10

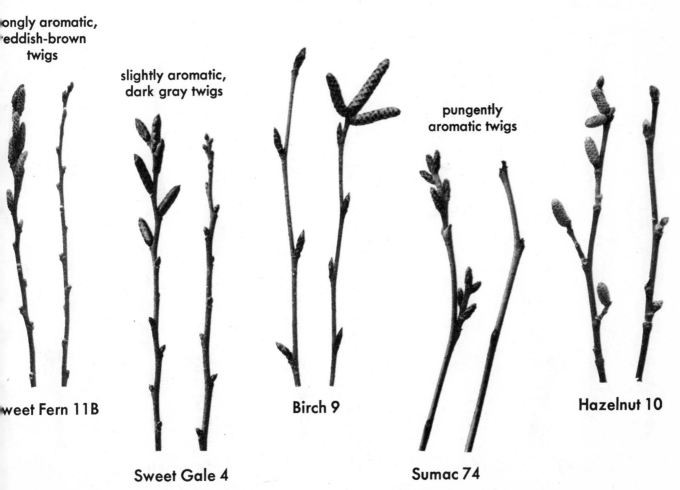

strongly aromatic,
reddish-brown
twigs

slightly aromatic,
dark gray twigs

pungently
aromatic twigs

TW
4

Sweet Fern 11B   Birch 9   Hazelnut 10

Sweet Gale 4   Sumac 74

Alternate Section

Note bud in *middle* of leaf-scar.

Note bud *above large* leaf-scar.

Note bud *above small* leaf-scar.

mostly opposit* but occasiona* partly alterna*

Sarsaparilla 92   Sumac 74   Sumac 74   Poison Sumac 74   Sumac 74   Smoke-tree 72   Fringe-tree 1

TW 5

flower bud ↓   leaf bud ↓

Note: All Willows have a single scale covering the bud. This comes off as a cap, or splits to allow the expanding growth to emerge. The "pussy" is the early stage of the Willow flower (see p. 4, Flower Key).

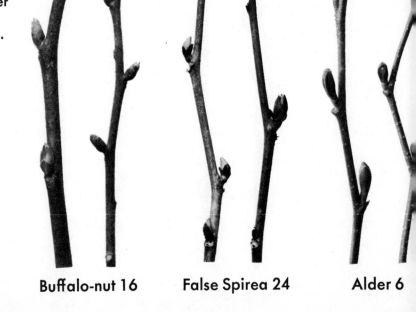

Willow 1   Buffalo-nut 16   False Spirea 24   Alder 6

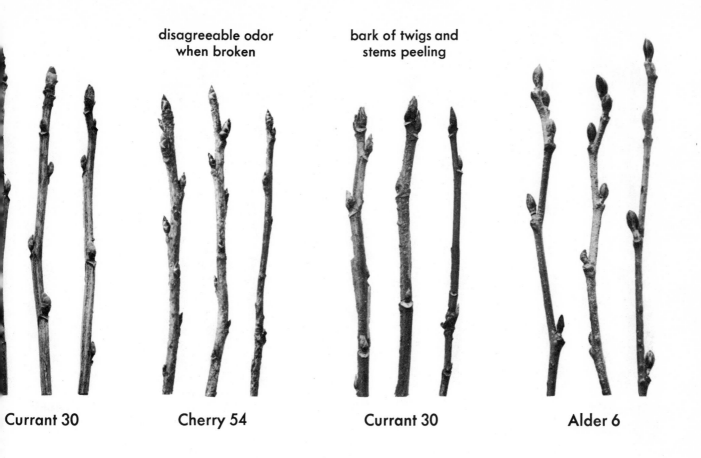

disagreeable odor
when broken

bark of twigs and
stems peeling

Currant 30   Cherry 54   Currant 30   Alder 6

inner bark of
twigs and stems
very yellow

red buds;
skunk odor when
broken

usually brown,
sometimes pink
buds

dark red
buds

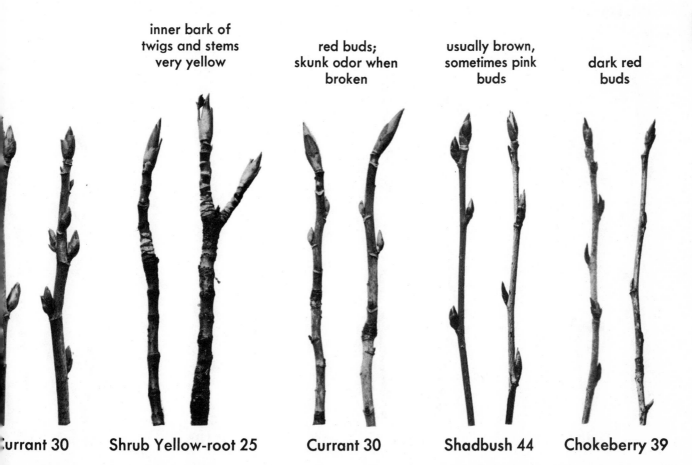

Currant 30   Shrub Yellow-root 25   Currant 30   Shadbush 44   Chokeberry 39

Daphne 91B   Alder 6   Silverbell 128   Alder 6

TW
7

twigs usually
hairy

tips of buds
whitish

red buds

Hazelnut 10   Alder 6   Hazelnut 10   Lyonia 117

red buds;
red and green twigs

buds usually
clustered at tips
of twigs

one scale
covers each bud

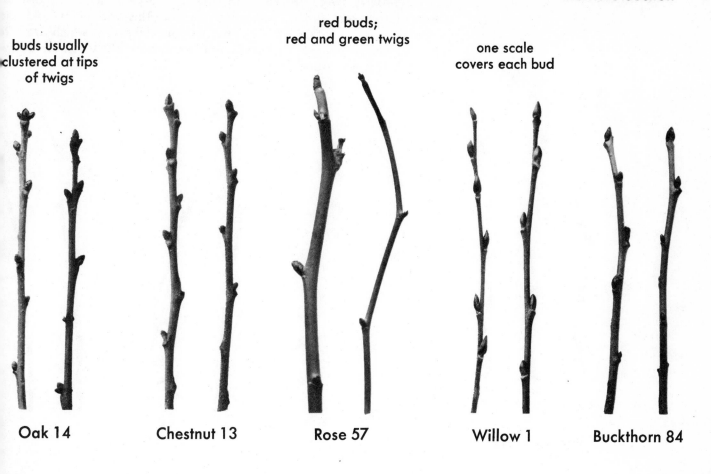

Oak 14   Chestnut 13   Rose 57   Willow 1   Buckthorn 84

TW
8

twigs somewhat
hairy (best seen
with lens)

often several buds
at each node (joint)

flower
buds

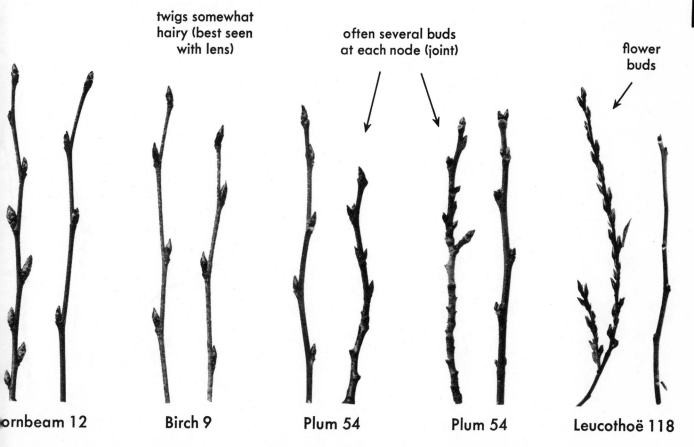

ornbeam 12   Birch 9   Plum 54   Plum 54   Leucothoë 118

Alternate Section

Note: Blueberry twigs are red and green. Two-year-old stems also a[re] often colored: red, green, yellow and brown. This variety of colorati[on] of the older wood is typical of Blueberries. Huckleberries, which are oft[en] confused with Blueberries, do not have this characteristic (see note belo[w]. (See Bark Key, pp. 2 and 18.)

flower buds   leaf buds

Blueberry 119

flower buds   leaf buds

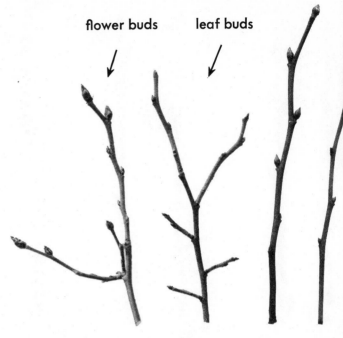

Blueberry 119

Note: Huckleberry twigs are typically red (sometimes pinkish, brown or dark gray). All other growth (two years and older) is uniformly dark gray (or brown) (see Blueberry note above). (See Bark Key, p. 16.)

very dark twigs, often with silvery coating

red buds; twigs and older stems (light) brown

leaf bud[s]

flower buds

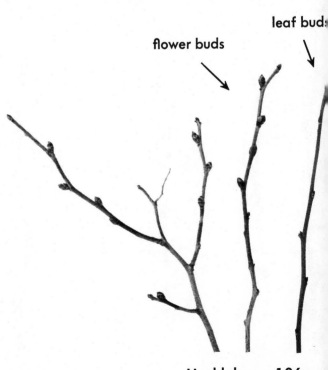

Holly 78   Lyonia 117   Huckleberry 126

124

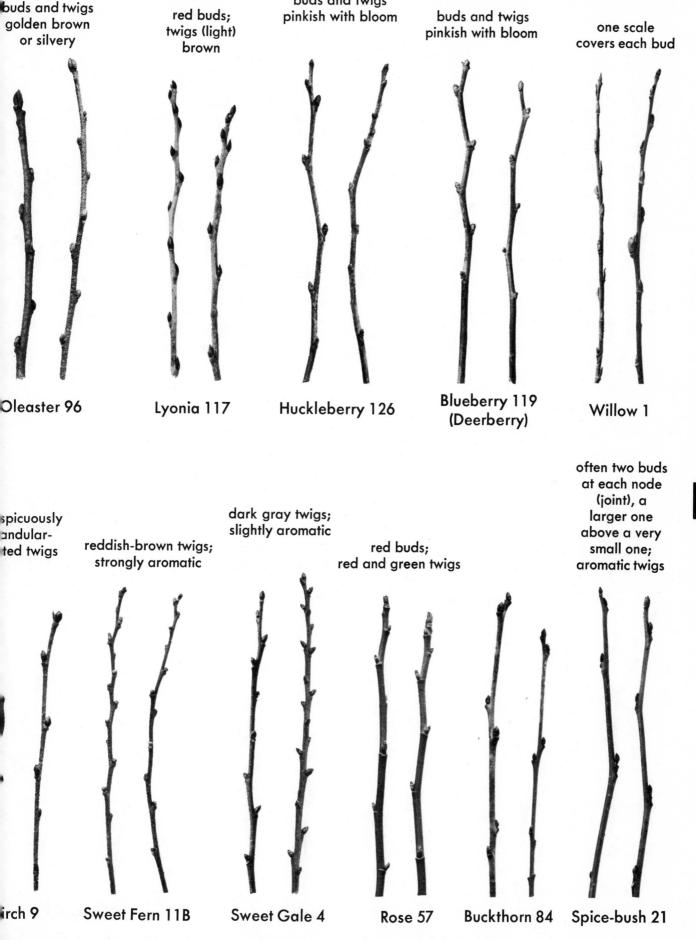

buds and twigs
golden brown
or silvery

red buds;
twigs (light)
brown

buds and twigs
pinkish with bloom

buds and twigs
pinkish with bloom

one scale
covers each bud

Oleaster 96   Lyonia 117   Huckleberry 126   Blueberry 119 (Deerberry)   Willow 1

often two buds
at each node
(joint), a
larger one
above a very
small one;
aromatic twigs

spicuously
andular-
ted twigs

reddish-brown twigs;
strongly aromatic

dark gray twigs;
slightly aromatic

red buds;
red and green twigs

irch 9   Sweet Fern 11B   Sweet Gale 4   Rose 57   Buckthorn 84   Spice-bush 21

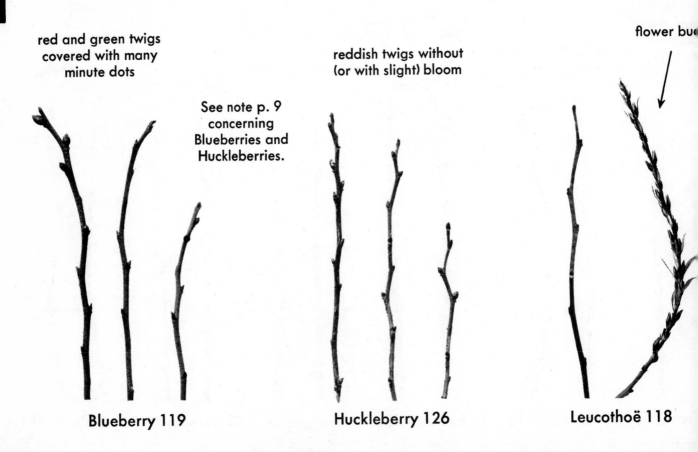

silvery,
fuzzy twigs

Blueberry 119
(Bilberry)

Blueberry 119
(Bilberry)

Oleaster 96

red and green twigs
covered with many
minute dots

See note p. 9
concerning
Blueberries and
Huckleberries.

reddish twigs without
(or with slight) bloom

flower bud

Blueberry 119

Huckleberry 126

Leucothoë 118

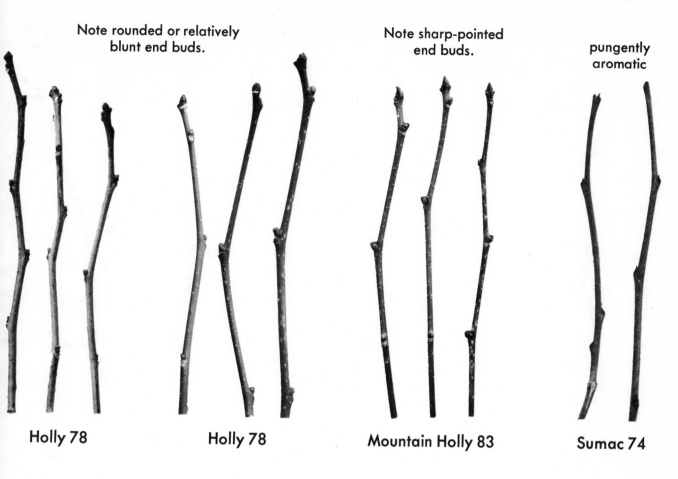

Note rounded or relatively blunt end buds.

Note sharp-pointed end buds.

pungently aromatic

Holly 78

Holly 78

Mountain Holly 83

Sumac 74

TW 12

Note: Large flower buds are on larger stems; only very small af buds are produced on twigs.

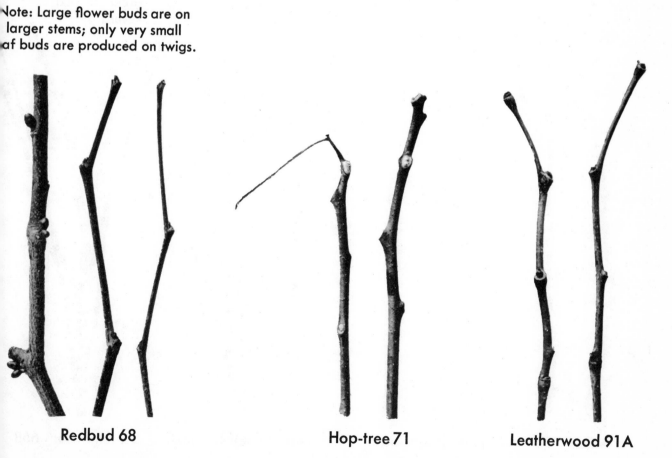

Redbud 68

Hop-tree 71

Leatherwood 91A

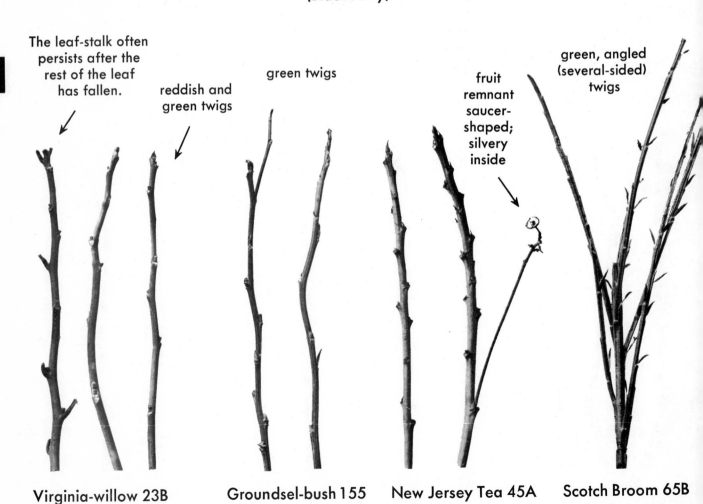

fruit remnants

twigs fuzzy;
buds shaggy

twigs red and green

buds round
and red; at
tips of twigs

Cinquefoil 64

Bramble 46
(Blackberry)

Hawthorn 43

TW
13

The leaf-stalk often
persists after the
rest of the leaf
has fallen.

reddish and
green twigs

green twigs

fruit
remnant
saucer-
shaped;
silvery
inside

green, angled
(several-sided)
twigs

Virginia-willow 23B

Groundsel-bush 155

New Jersey Tea 45A

Scotch Broom 65B

fruit remnants

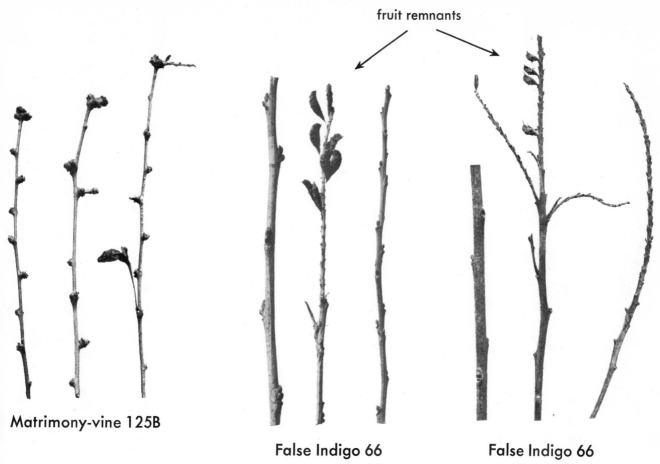

Matrimony-vine 125B

False Indigo 66

False Indigo 66

green, angled (several-sided) twigs

twigs silvery, fuzzy

Dyer's Greenweed 65A

Oleaster 96

Rose of Sharon 90

The twigs shown on this page
largely die back in winter, but buds
can usually be found on larger stems.

False Spirea 24

Ninebark 27

**Bramble 46
(Raspberry)**

TW
15

mostly
opposite,
but upper
twigs with
alternate
characteristics

Marsh-elder 156

Spirea 40

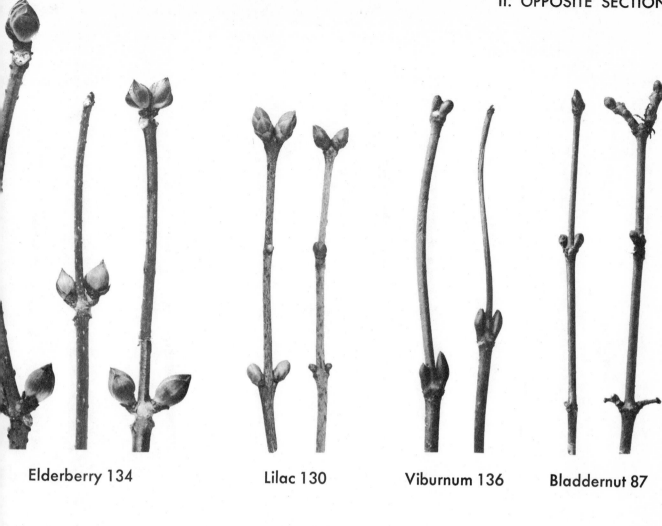

Elderberry 134   Lilac 130   Viburnum 136   Bladdernut 87

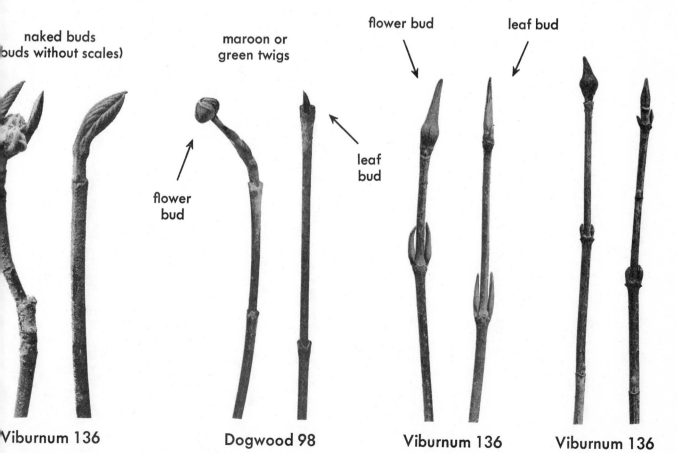

naked buds
(buds without scales)

maroon or
green twigs

flower bud   leaf bud

flower
bud

leaf
bud

Viburnum 136   Dogwood 98   Viburnum 136   Viburnum 136

Note: Side buds in the top row are markedly appressed (close to ste

growth (leaf) buds

flower bud

pinkish or
yellow twigs

bright red twig
(yellow in variet

Viburnum 136

Dogwood 98

Viburnum 136

Dogwood

Forsythia 131

Viburnum 136

Honeysuckle 146

Elderberry 13

opposite, but
often partly
alternate

red, maroon or
green twigs

Buckthorn 84   Viburnum 136   Dogwood 98   Viburnum 136

opposite, but sometimes
partly alternate

this type
of growth
sometimes,
but not
always,
found

Hydrangea 28   Fringe-tree 129   Honeysuckle 146

fruit (or remnants)
present most of year

twigs and buds
usually silvery

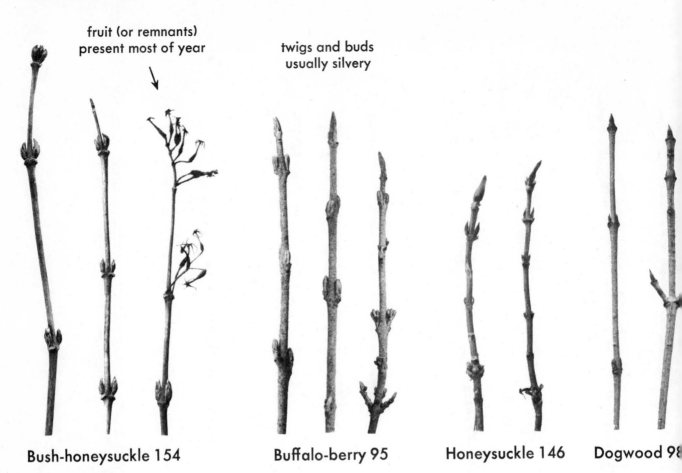

Bush-honeysuckle 154   Buffalo-berry 95   Honeysuckle 146   Dogwood 98

green or maroon twigs;
often four-sided

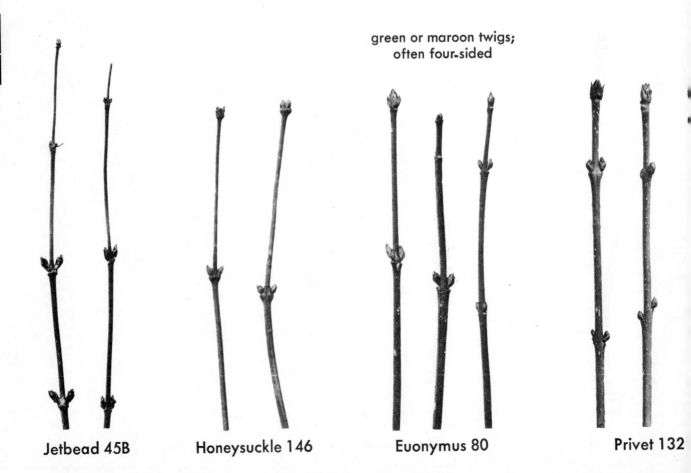

Jetbead 45B   Honeysuckle 146   Euonymus 80   Privet 132

Opposite Section

twigs with slight ridge
(or small wing) along
opposite sides

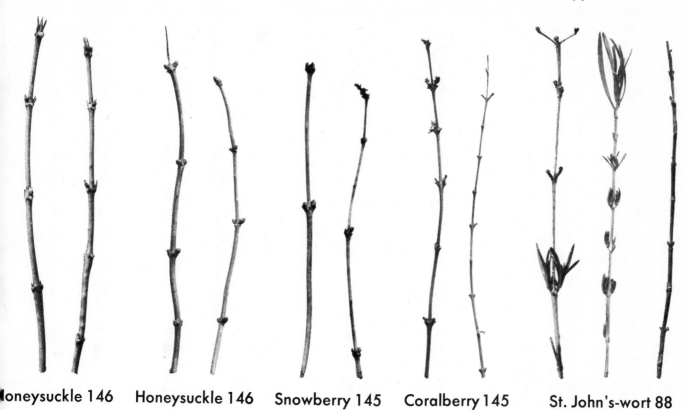

Honeysuckle 146   Honeysuckle 146   Snowberry 145   Coralberry 145   St. John's-wort 88

green or maroon
twigs

fuzzy twigs

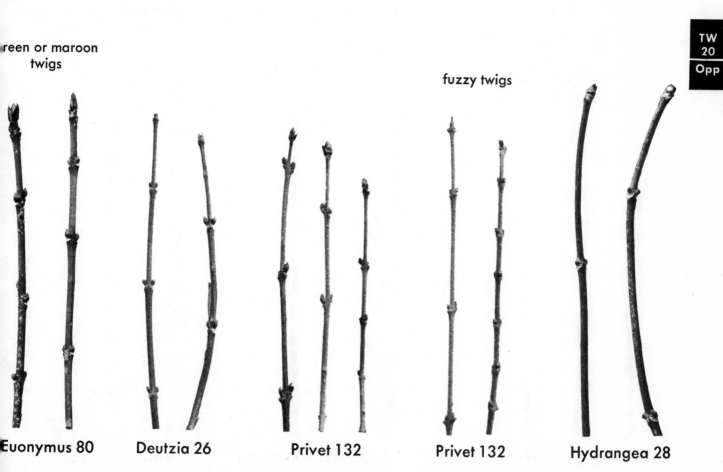

Euonymus 80   Deutzia 26   Privet 132   Privet 132   Hydrangea 28

Opposite Section

aromatic twigs

maroon and
green twigs

buds often in whorls of
three or four around
stems at each node
(joint)

Sweet Shrub 20

Mock-orange 22

Dogwood 98

Buttonbush 144

mostly opposite, but upper
twigs with alternate
characteristics

large, distinctive wings
(see also Bark Key, p. 23)

distinctive wings alternating
at right angles along twigs

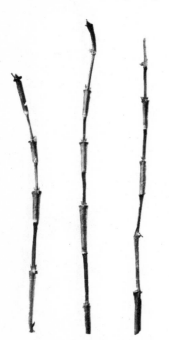

St. Peter's-wort 89A

Euonymus 80

Marsh-elder 156

Key #6—BARK

This Key is divided into two main sections (determinable at any time of the year):

I ALTERNATE: barks of shrubs having alternate characteristics (leaves, buds and leaf-scars), pp. 1-18

II OPPOSITE: barks of shrubs having opposite characteristics, pp. 19-27

Within the above sections, similar barks are grouped together, such as shaggy, striped, spotted, nondescript, or relatively smooth barks. It will be apparent from a perusal of this Key and an examination of actual shrubs that no exact grouping is possible. Bark is not a detail which is as definite and positively characteristic as the details shown in other Keys, although many barks are truly distinctive. The Bark Key should be used last to verify identification made in previous Keys and, used in this way, will prove invaluable, often being the deciding factor in questions of doubt. Used first, it will only prove confusing. However, unlike the other details, barks are always present.

Two (or more) barks of each plant are always shown together. The bark of larger stems of some shrubs is quite different from that of the smaller (or younger) growth, while on other shrubs the bark is consistently uniform throughout. Therefore always check both sizes, the larger one first, to determine the position in the Key. In selecting the barks for this Key a definite size limit was established for the larger barks. After a certain size, most barks become very roughened and are less distinctive than on the smaller and younger stems. Therefore, in comparing barks of actual shrubs with the pictures shown here, keep within the sizes indicated. Actually, stems of the correct size will prove to be much more accessible than the larger ones, concealed in the middle of a shrub which is often almost impenetrable.

Some shrubs commonly have two or more quite different kinds of bark on different plants of the same species. There are a number of reasons for this. Sometimes it is a question of location or environment, but these differences may also be seen on plants apparently growing under similar conditions. In such cases, a variety may be indicated or possibly a particular species is not a stable one and is still undergoing evolutionary changes. Occasionally, the bark of the male (staminate) plant is decidedly different from that of the female (pistillate) plant. In any event, regardless of cause, if a consistent variation is commonly encountered, both types of bark will be found in appropriate places in this Key. (*Note:* No attempt has been made to explain the individual differences shown here, as this is a little-explored subject outside the scope of this book. But it would be an interesting one to pursue. It might even lead to important discoveries concerning the nature of plants and their evolution.)

Look at both sides of the stem, as bark in the sun is often different from that on the shaded side. One last word of caution: Do not try to guess what a bark covered by lichens, fungus or moss really looks like; find a section not covered over, or remove the growth carefully.

I. ALTERNATE SECTION

The plants on these two pages are all closely related. Their bark although not identical, have much in common. It is well to becon familiar with this type of bark, as it distinguishes these plants fro others in this section.

  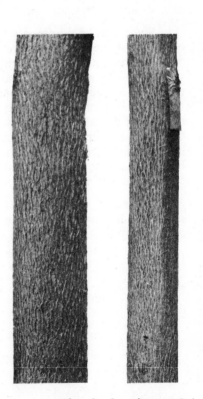

Laurel 110

Andromeda 114

Rhododendron 104

Rhododendron 104

green, red
or maroon

pale
brown

Blueberry 119

Clethra 102

BK
2

Lyonia 117

Lyonia 117

Azalea 104

Ninebark 27

Cinquefoil 64

Oleaster 96

Oak 14

small stems
green or maroon

Bramble 46
(Raspberry)

Sand-myrtle 113B

Leucothoë 118

Smoke-tree 72

Hawthorn 43

BK
4

green

Matrimony-vine 125B

Groundsel-bush 155

Scotch Broom 65B

Chokeberry 39

Plum 54

Cherry 54

Silverbell 128   Shadbush 44   Currant 30

Dogwood 98
(Alternate-leaved)   Rose 57

BK
6

Alternate Section

small stems
red or maroon

aromatic bark

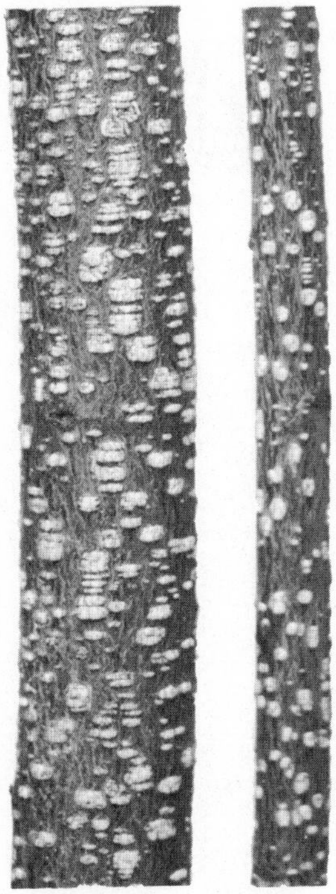

Buckthorn 84

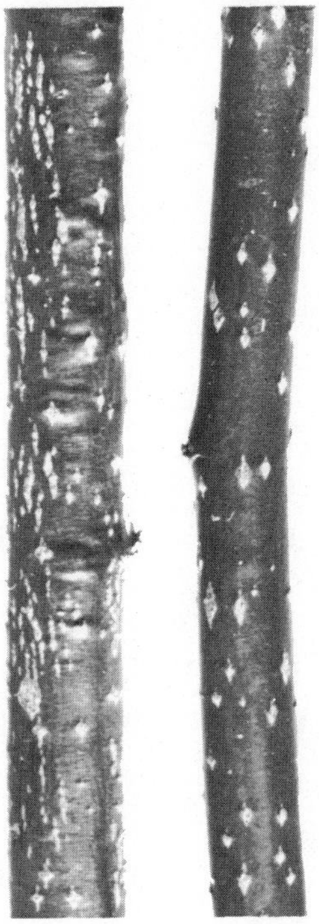

Dogwood 98
(Alternate-leaved)

Sweet Fern 11B

False Spirea 24

Beware of this pla▮
it is worse than Pois▮
Ivy for many peop▮

aromatic bark

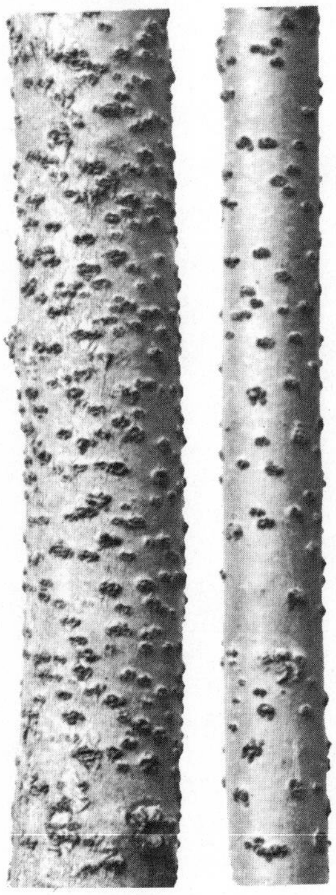

Sumac 74

Currant 30

Spice-bush 21

Poison Sumac 7▮

aromatic bark

Bayberry 4

Bristly Locust 69

Firethorn 42

mostly opposite, but
ften partly alternate

BK
8

Fringe-tree 129

Sumac 74

Birch 9

Alder 6

145

lenticels (dots on bark)
usually orange

large, soft
pith

small ste[m]
often hai[r]

Alder 6

Holly 78

Sumac 74

large, soft pith

large, soft pith

BK
9

Sumac 74

Sumac 74

Hop-tree 71

aromatic bark

| Mountain Holly 83 | Holly 78 | Sweet Gale 4 |

aromatic bark

| Oak 14 | Oleaster 96 | Wax-myrtle 4 |

BK
10

mostly opposite, but
often partly alternate

Chokeberry 39

Oleaster 96

Buckthorn 84

Plum 54

Buffalo-nut 16

Sarsaparilla 92

BK
11

aromatic bark

Willow 1

Spice-bush 21

Alder 6

Holly 78

Chestnut 13
(Chinquapin)

Oak 14

BK
12

red-brown bark

Currant and Gooseberry 30   **Hazelnut 10**   **Hazelnut 10**

red-brown bark

BK
13

Buckthorn 84   **Hornbeam 12**   **Witch-Hazel 38**

Beware of this plant; it is worse than Poison Ivy for many people. It is found almost entirely in swamps and wet ground.

pungently aromatic bark

Poison Sumac 74

Smoke-tree 72

Fragrant Sumac 74

bark very strong and pliable; wood soft and weak

small stems green, tinged with red

BK 14

Leatherwood 91A

Virginia-willow 23B

Rose of Sharon 90

inner bark yellow

Daphne 91B

Shrub Yellow-root 25

Azalea 104

Note flower
buds on
large stems.

sm
ste
gre

Redbud 68

New Jersey Tea 45A

Dyer's Greenweed 65A

BK
15

Buckthorn 84   **False Indigo 66**   **Labrador Tea 116A**   **Leatherleaf 116B**

BK
16

alse Indigo 66   **Willow 1**   **Leucothoë 118**   **Huckleberry 126**

Alder 6

Clethra 102

Azalea 104

stems rust-colored; flaky bark

stems reddish
brown; flaky ba

Hornbeam 12

Menziesia 103

Currant 30

stems bamboo-like; hollow

stems red, maroon or green; solid pith

stems golden brown or red (sometimes with bloom)

stems reddish brown; skunk odor when bruised

notweed 17B

Bramble 46
(Blackberry)

Bramble 46
(Raspberry)

Currant 30

small stems green, red, maroon, yellow or brown

old stems mostly light gray

young growth brown or golden brown

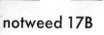

Blueberry 119

Spirea 40

BK 18

Deutzia 26

St. John's-wort 88

St. Peter's-wort 8

Buttonbush 144

Coralberry 145

Hydrangea 28

Honeysuckle 146   Bush-honeysuckle 154   Mock-orange 22

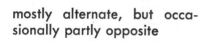

mostly alternate, but occasionally partly opposite

actually alternate, but leaves always clustered at ends of stems, giving a definite opposite appearance

Laurel 110   Andromeda 114   Rhododendron 104

Snowberry 145 **Hydrangea 28** **Viburnum 136**

Lilac 130 **Buffalo-berry 9**

Dogwood 98

Dogwood 98

Viburnum 136

Viburnum 136

BK
22
Opp

vigorous young shoots green
with four light tan stripes

Euonymus 80

Honeysuckle 146

Euonymus 80

stems (small ones especially)
greenish with dull orange stripes

small stems red, maroon or gre

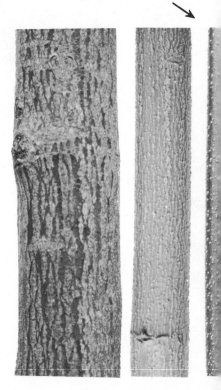

BK
23
Opp

Bladdernut 87

Euonymus 80

Dogwood 98

large pith

Elderberry 134   **Viburnum 136**   **Buckthorn 84**

large pith

Buffalo-berry 95   **Viburnum 136**   **Elderberry 134**

aromatic bark

Jetbead 45B

Fringe-tree 129

Sweet Shrub 20

large and small stems red

large and small stems
pinkish or greenish yellc

Forsythia 131

Dogwood 98

Dogwood 98

BK
25

Opp

Privet 132

Viburnum 136

Euonymus 80

Privet 132

Viburnum 136

Privet 132

BK
26
Opp

Viburnum 136

Dogwood 98

green

Viburnum 136

Marsh-elder 156

Mistletoe 17A

PART II

MASTER PAGES

(all actual size except when noted)

MASTER PAGES

The Master Pages have two main uses: (1) to assemble in one place the important features of each shrub, and (2) to identify species.

At the heading of each Master Page the common name is given, followed by the botanical one. This is followed by a flower note: whether *perfect* (both male and female parts in the same flower), *monoecious* (distinct male and female flowers on the same shrub), or *dioecious* (male flowers on one plant, female flowers on a different one). The approximate date of blooming is given. As the dates vary somewhat in different years (especially for the early-bloomers) and as the same shrubs bloom at different times in different parts of their range, no exact date can be given. Comparisons with a few known dates of bloom will indicate more exactly what to expect in any particular section of the country.

Note is made as to whether the shrubs are alternate or opposite. A description of range is given, but it should be realized that the outside limits indicated should not be taken to mean that any particular shrub will be found throughout the area. Some plants grow only at high elevations within their indicated range, others only at low altitudes; some only in wet soil, others only in dry soil, etc.

The heights given for each shrub are for average mature plants, and, of course, occasional giants can be found of almost any plant. However, a plant indicated as 2-3 feet will rarely be found 5 feet tall, and the plants that are normally very large when full grown will certainly show evidence of this at an early age.

The headings are followed by the pictures of actual details, and in cases where several species are included in one genus, comparison of these details leads to species identification. As explained before, it is by comparing all available details that final identification is achieved, and the combination of determining factors is not the same in each case, depending on the season or the characteristics of the plant itself. Throughout the Master Pages, notes point out differences among species, when identification might be difficult otherwise.

WILLOW — *Salix*   Dioecious   Alternate (rarely sub-opposite); few foreign species

Beak Willow *Salix bebbiana*	late April-early June	Nfd. to Alaska, s. to Md., Ill., Ia. & west	25 ft.
Bearberry Willow *S. uva-ursi*	June-July	Lab. to Alaska, s. to alpine summits of N.H. & N.Y.	Prostrate
Dwarf Gray Willow *S. tristis* (Dwarf Upland W.; Dwarf Pussy W.) (Considered by some botanists as a variety of Gray Willow.)	March-April	Me. to Minn., s. to Fla. & Okla.	1½ ft.
Gray Willow *S. humilis* (Upland W.; Prairie W.)	March-early June	Nfd. to Minn., s. to N.C. & Kans.	8 ft.
Pussy Willow *S. discolor* (Glaucous W.)	late Feb.-May	N.S. to Man., s. to Va., Mo. & west	20-25 ft.
Shining Willow *S. lucida*	late April-June	Lab. to Man., s. to N.J. & Ia.	20 ft.
Silky Willow *S. sericea*	late March-early May	Que. to Mich., s. to S.C. & Mo.	12 (occ. 15) ft.

There are a great many named species of Willows, Rehder's manual listing sixty-three, and Gray's manual fifty-four. Both manuals list many more hybrids and forms, so that the total number is truly impressive. However, as explained in the Introduction to this book, certain genera contain plants that do not consistently remain true to any species type. The Willows are such a genus, and therefore only a few species are given here to indicate typical forms and the means by which these species, if found true to type, can be identified.

The method used here for the Willows is based primarily on summer characteristics. Winter identification of species is difficult and should not be attempted at first. It is interesting to note, however, that the detail most helpful in distinguishing any Willow from other plants is often considered a winter characteristic: the single bud scale that covers the individual buds of all Willows. All other shrubs included in this book have two or more scales on each bud, or have naked buds (buds without scales). Although buds are more evident in winter, they can usually be found on most plants from midsummer on. A winter-killed bud good enough for this purpose can often be found in spring. Thus one can identify a Willow as such at almost any time of year, and this is usually sufficient for this difficult genus.

MP
1

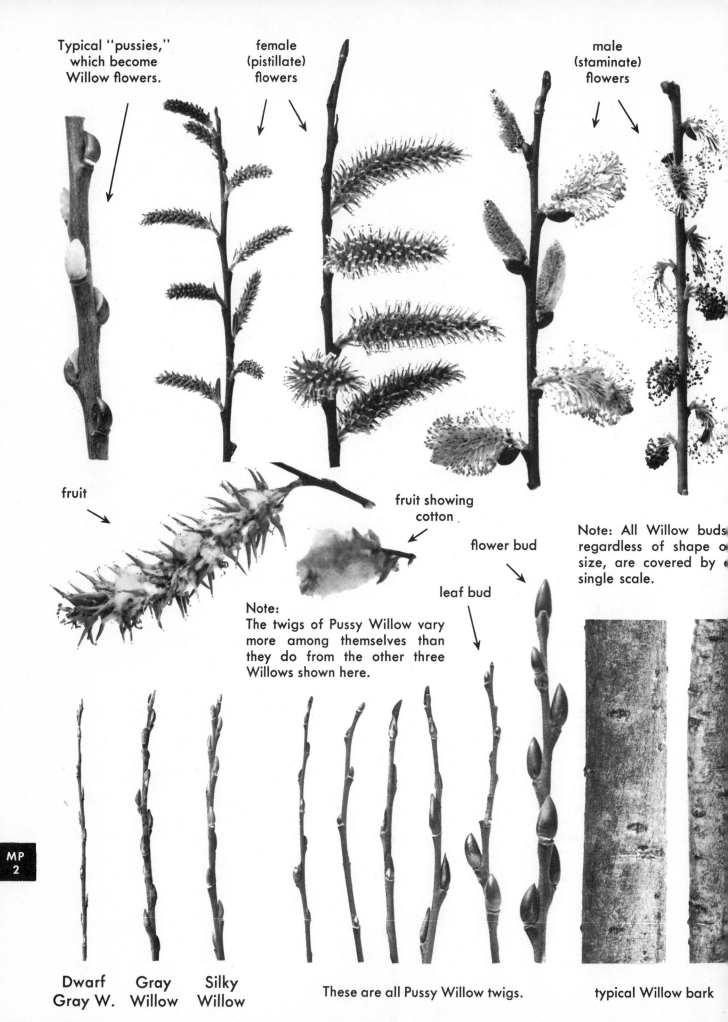

Typical "pussies," which become Willow flowers.

female (pistillate) flowers

male (staminate) flowers

fruit

fruit showing cotton

Note: All Willow buds regardless of shape or size, are covered by a single scale.

flower bud

leaf bud

Note:
The twigs of Pussy Willow vary more among themselves than they do from the other three Willows shown here.

MP
2

Dwarf   Gray   Silky
Gray W.   Willow   Willow

These are all Pussy Willow twigs.

typical Willow bark

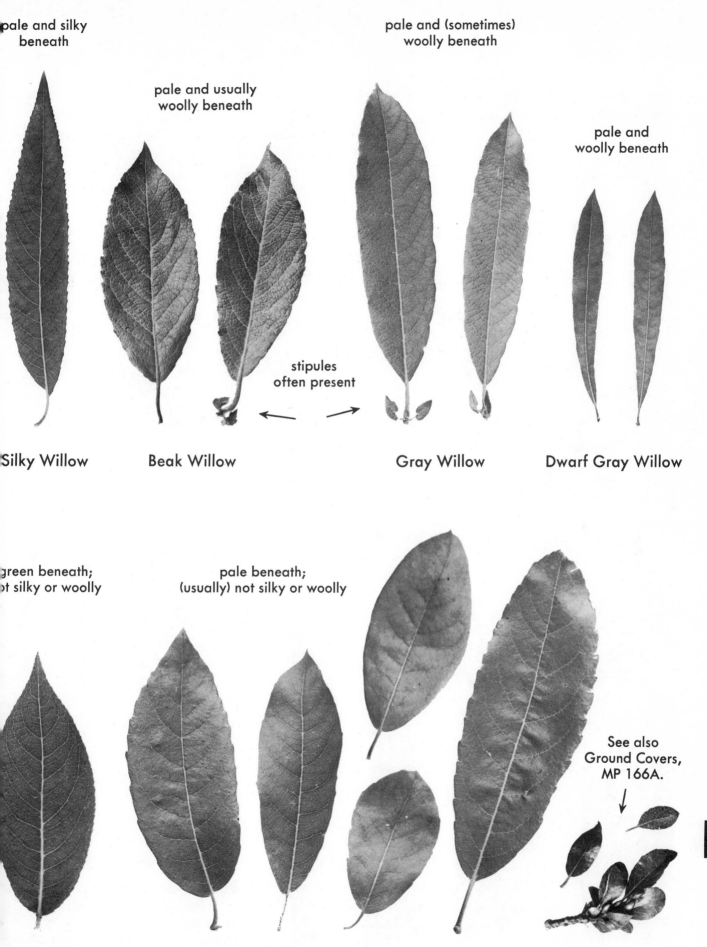

pale and silky
beneath

pale and usually
woolly beneath

pale and (sometimes)
woolly beneath

pale and
woolly beneath

stipules
often present

Silky Willow

Beak Willow

Gray Willow

Dwarf Gray Willow

green beneath;
not silky or woolly

pale beneath;
(usually) not silky or woolly

See also
Ground Covers,
MP 166A.

Shining Willow

These five leaves are all Pussy Willow
(all five forms are often found on the same plant).

Bearberry Willow

MP
3

169

BAYBERRY; GALE; WAX-MYRTLE—*Myrica*

Dioecious or Monoecious Alternate shrubs with aromatic leaves, twigs and bark

Bayberry	*Myrica pensylvanica*	May-July	Nfd. to N. C., largely on coastal plain; locally w. to n. Ohio and w. N. Y. near Lake Erie 10
Evergreen Bayberry (Wax-myrtle)	*M. heterophylla*	April-June	coastal plain, s. N. J. to Fla. & La. 10 (occ. 15
Sweet Gale	*M. gale*	April-June	N. Amer.: s. to s. N. E., s. N. Y., n. Pa., Mich., Wis., Minn. & west; also occ. s. in mts. to N. C. & Tenn. 5-(
Wax-myrtle	*M. cerifera*	April-June	s. N. J. mostly near coast to Fla. & Tex. 3.

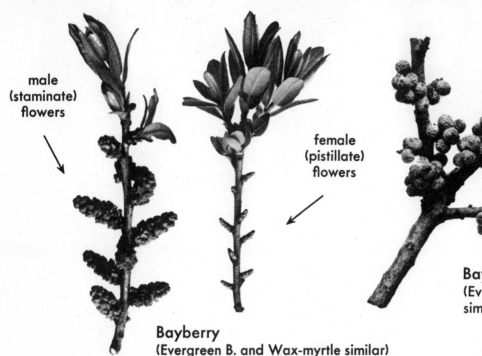

male (staminate) flowers

female (pistillate) flowers

Bayberry
(Evergreen B. and Wax-myrtle similar)

waxy, berry-like fruit

Bayberry
(Evergreen B. and Wax-myrtle similar but slightly smaller)

barks typical of Bayberry, Evergreen B. and Wax-myrtle

Bayberry: warm brown, tan or light gray
Wax-myrtle: similar
Evergreen Bayberry: similar shape but densely fuzzy and usually blackish green in winter

MP
4

Aromatic leaves, all with typical Bayberry odor

evergreen (semi-evergreen
in n. part of range)
small golden-yellow resin dots
on *both* sides (best seen with lens)
narrow leaves

evergreen
resin dots beneath only
pale and dull beneath

deciduous (sometimes lasting
into winter)
resin dots beneath only
yellow green and shiny beneath

Wax-myrtle   **Evergreen Bayberry**   **Bayberry**

winter
catkins

golden-brown
male (staminate)
flowers

reddish
female
(pistillate)
flowers

fruit

Sweet Gale

Sweet Gale is considered a subgenus. The leaves are similar to the Bayberries and Wax-myrtle, and all are aromatic shrubs, but notice that Sweet Gale is the only one with winter catkins and that its flowers, fruit, and twigs are quite different from the others.

ALDER—*Alnus*   Monoecious   (Flowering dates are important)   Alternate

Mountain Alder   *Alnus crispa*   June-July   Lab. to Alaska, s. to Mass. & w. to Minn.;
    (American Green Alder)   also appears in mts. of w. N. C.   10
Sea-side Alder   *A. maritima*   Aug.-Sept.   Del. & Md., also Okla.   30-35
Smooth Alder   *A. rugosa*   Feb.-May   N. S. & Me., N. Y., Ohio, Ind., Ill., Mo. &
            Okla., s. to Fla. & La.   15 (occ. 25)

Speckled Alder   *A. incana*   March-May   N. S. to Minn., s. to s. N. E., w. to W.
            Va., n. Ill. & n.e. Ia.   15 (occ. 25)

Note: Both the Smooth and Speckled Alders grow side by side in much of their range. Hybridizatio
occurs frequently, making identification confusing. However, when true to type, the differences a
evident; when not, it is best to suspect a hybrid, or intermediate form. The other two Alders shown he
are easy to identify at any time of year.

female
(pistillate)
flowers

developing into fruit (seed-bearing cones)

conspicuous thin
wings on seed

← seeds x 3 →

seed without wing

Mountain Alder   other Alders

male
(staminate)
flowers

typical
Alder flowers

Mountain Alder   Sea-side Alder   Smooth Alder
                            (Speckled A. simila

Mountain Alder   Sea-side Alder   Smooth Alder   Speckled Alder

MP
6

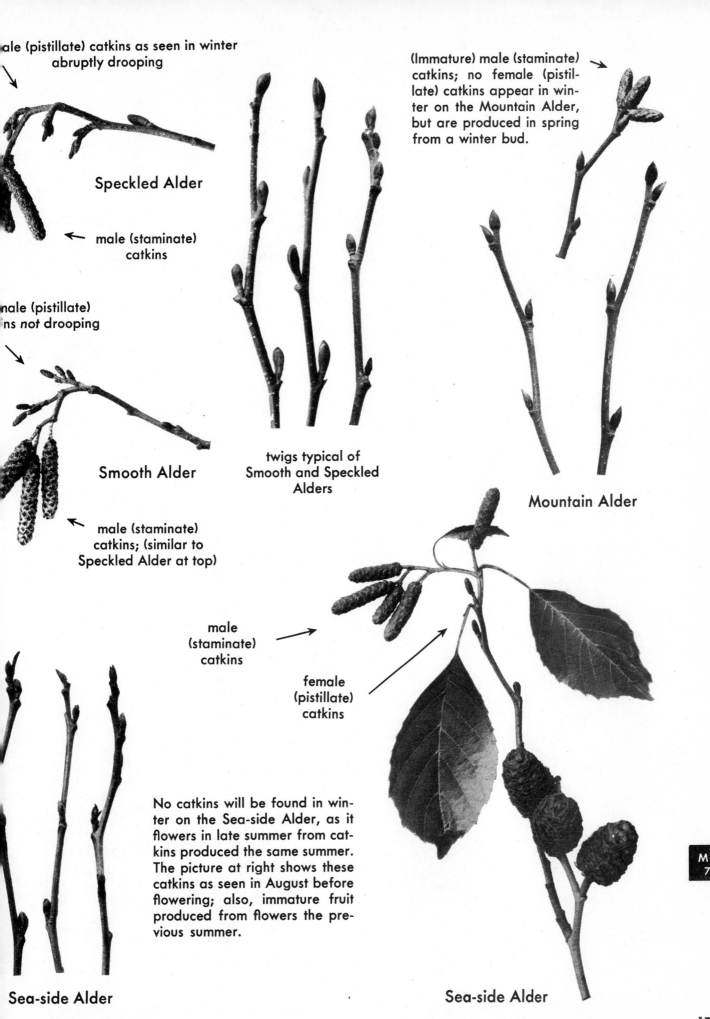

ale (pistillate) catkins as seen in winter abruptly drooping

Speckled Alder

← male (staminate) catkins

(Immature) male (staminate) catkins; no female (pistillate) catkins appear in winter on the Mountain Alder, but are produced in spring from a winter bud. →

ale (pistillate) ns *not* drooping

Smooth Alder

twigs typical of Smooth and Speckled Alders

Mountain Alder

← male (staminate) catkins; (similar to Speckled Alder at top)

male (staminate) catkins →

female (pistillate) catkins

No catkins will be found in winter on the Sea-side Alder, as it flowers in late summer from catkins produced the same summer. The picture at right shows these catkins as seen in August before flowering; also, immature fruit produced from flowers the previous summer.

Sea-side Alder

Sea-side Alder

MP
7

173

Mountain Alder

Speckled Alder

Sea-side Alder

Smooth Alder

MP
8

BIRCH—*Betula* Monoecious Alternate

Dwarf Birch	*Betula glandulosa*	June-Aug.	Nfd. to Alaska, s. to mts. of Me., N. H. & N. Y.; also appears in Mich. & some w. states 6 ft.
Low Birch	*B. pumila*	May-June	Nfd. & west, s. to Conn., Ill. & Ia. 15 ft.

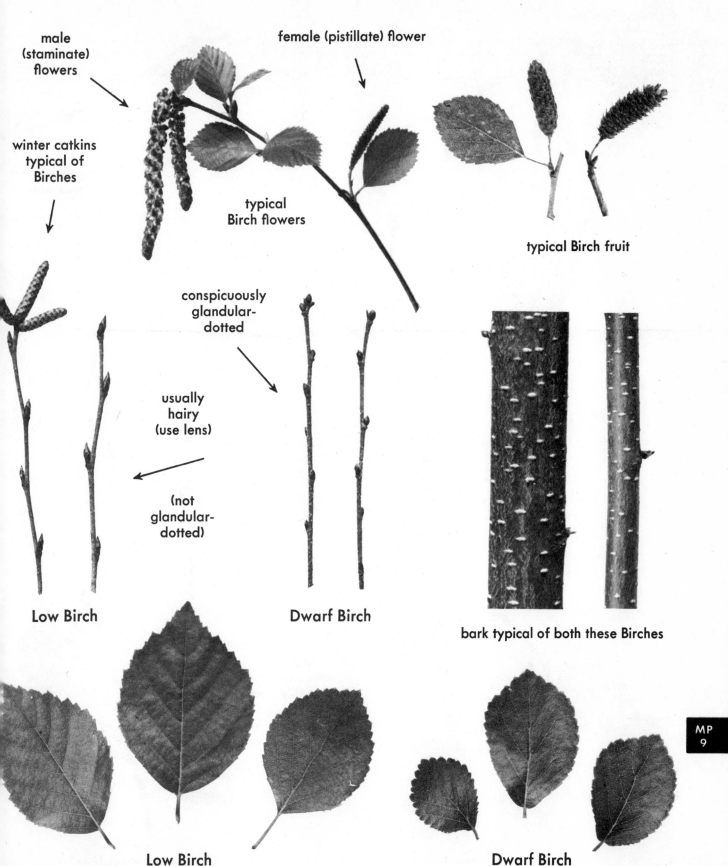

male (staminate) flowers

female (pistillate) flower

winter catkins typical of Birches

typical Birch flowers

typical Birch fruit

conspicuously glandular-dotted

usually hairy (use lens)

(not glandular-dotted)

Low Birch

Dwarf Birch

bark typical of both these Birches

Low Birch

Dwarf Birch

MP 9

HAZELNUT or HAZEL—*Corylus* Monoecious Alternate

American Hazelnut	*Corylus americana*	April-May	Me. to Sask., s. to Ga. & Okla. 10
Beaked Hazelnut	*C. cornuta*	April-May	Nfd. to B. C., s. to Ga., Mo. & Colo. 10

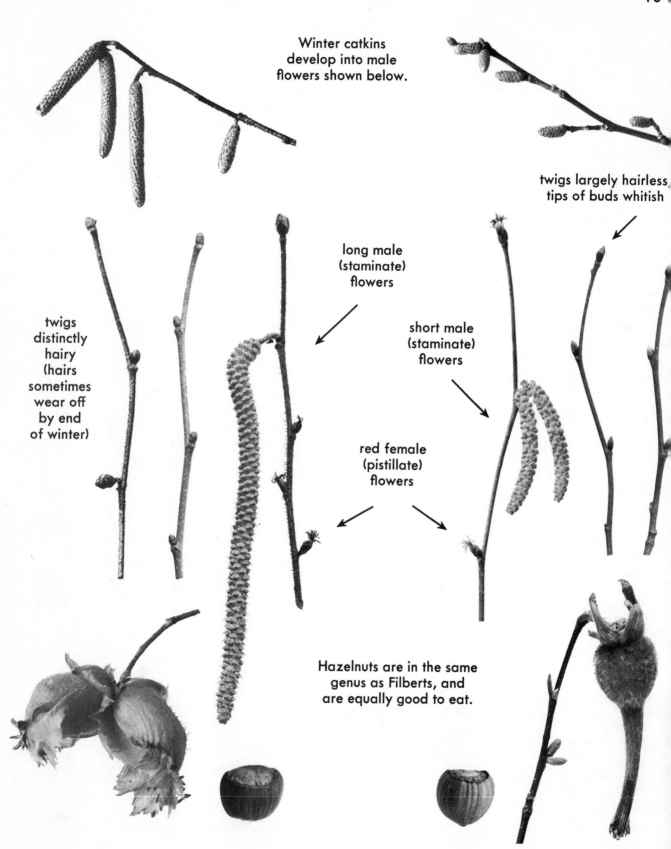

Winter catkins develop into male flowers shown below.

twigs largely hairless, tips of buds whitish

long male (staminate) flowers

short male (staminate) flowers

twigs distinctly hairy (hairs sometimes wear off by end of winter)

red female (pistillate) flowers

Hazelnuts are in the same genus as Filberts, and are equally good to eat.

American Hazelnut

Beaked Hazelnut

eaf-stalk
fuzzy ➝

leaf-stalk
not fuzzy ⬅

typical Hazelnut barks

American Hazelnut

Beaked Hazelnut

SWEET FERN—*Comptonia peregrina* Monoecious   April-May   Alternate
Lab. to Alaska, s. to N. C., Tenn., Ill., Wis., Minn. & west 5 ft.

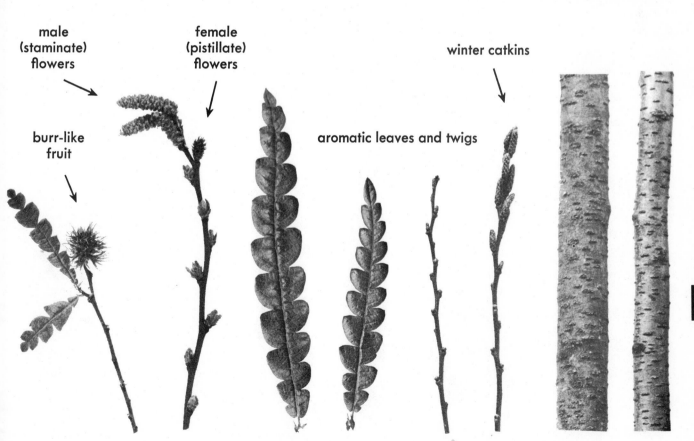

male
(staminate)
flowers ➝

female
(pistillate)
flowers ➝

winter catkins ➝

burr-like
fruit ➝

aromatic leaves and twigs

MP
11

AMERICAN HORNBEAM—*Carpinus caroliniana* Monoecious April-May   Alternate
(Blue Beech; Ironwood)
N.S. to Minn., s. to Fla. & Tex.   35 ft.

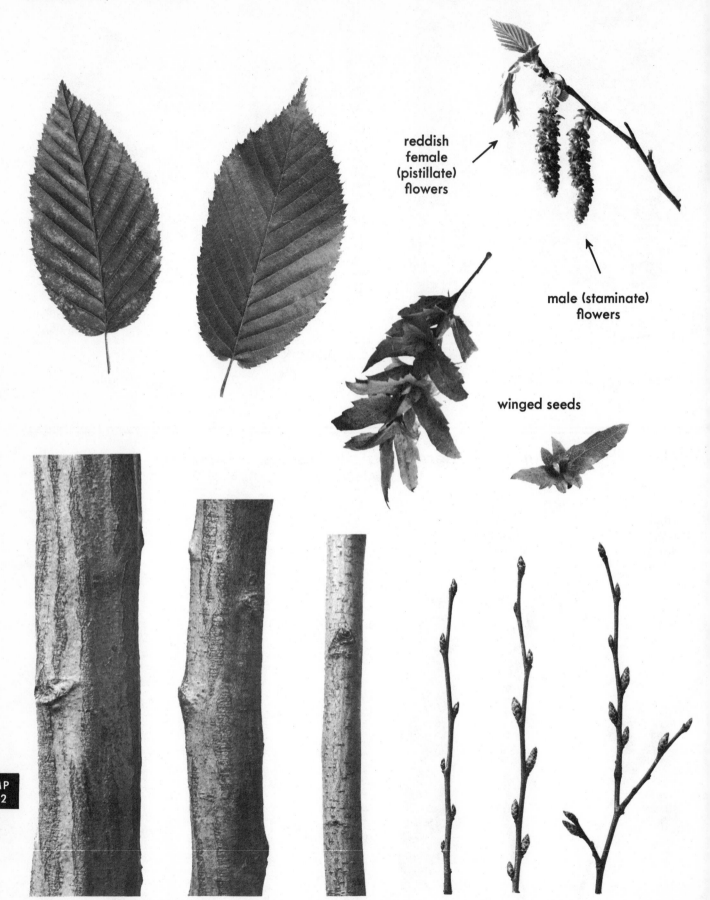

reddish female (pistillate) flowers

male (staminate) flowers

winged seeds

typical muscular-looking stems

CHESTNUT—*Castanea* Monoecious Alternate

Chinquapin *Castanea pumila* June s. N. Y., e. Pa., W. Va., Tenn., s. to Fla. & Tex.

15-30 ft.

female
(pistillate)
flowers

male
(staminate)
flowers

Only a single nut is
produced in each burr.

OAK—*Quercus*   Monoecious   Alternate

Chinquapin Oak	*Quercus prinoides*	April-May	Me. to Minn., s. to Ala. & Tex.
White Oak Group			10 (occ. 15)
Scrub Oak	*Q. ilicifolia*	April-May	Me. to N. Y., s. to N. C. & W. Va.
Red (or Black) Oak Group			18-20

Note: There are two main subgenera: (1) the White Oak Group, with acorns that mature in one year and leaves wit[h] lobes that may be rounded or sharp-pointed but never bristle-tipped; (2) the Red (or Black) Oak Group, with acorns tha[t] require two years to mature and leaves with bristle-tipped lobes. Considerable hybridization occurs among differe[nt] species within the same group, and this must be borne in mind when identifying Oaks, as they will not always appea[r] true to type.

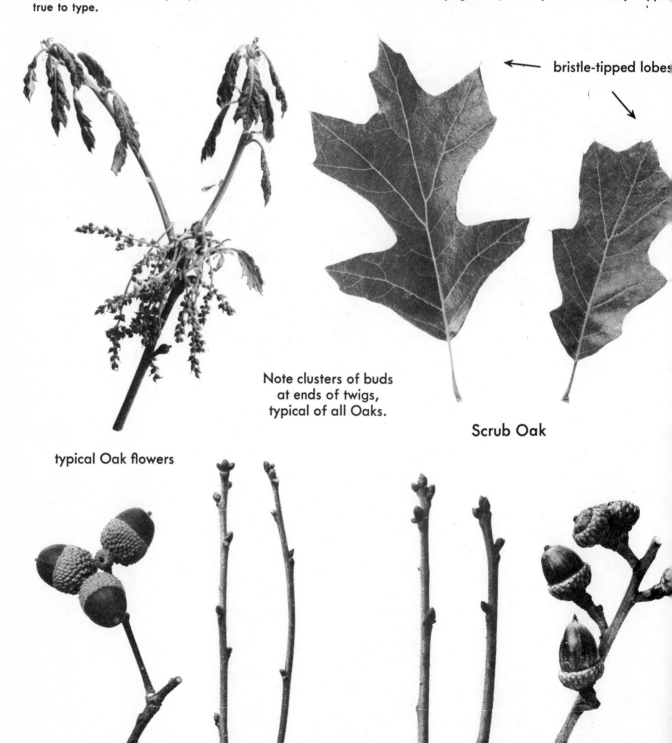

← bristle-tipped lobes

Note clusters of buds
at ends of twigs,
typical of all Oaks.

Scrub Oak

typical Oak flowers

Chinquapin Oak

Scrub Oak

bristle-tipped lobes

lobes (teeth) usually sharp-pointed
but *not* bristle-tipped

Scrub Oak

Chinquapin Oak

Scrub Oak

Chinquapin Oak

MP
·15

181

BUFFALO-NUT—*Pyrularia pubera*
(Oil-nut)

The Buffalo-nut is parasitic on the roots of other shrubs and trees, but appears to be a separate shrub.

Note: The twigs fall off easily, leaving round twig - scars above the leaf-scars.

twig-scar

leaf-scar

MP
16

182

MERICAN MISTLETOE— *Phoradendron flavescens* Dioecious Oct.-Dec. Opposite
N. J. to Kans., s to Fla. & Tex.; evergreen;
parasitic on branches of various trees   3 ft.

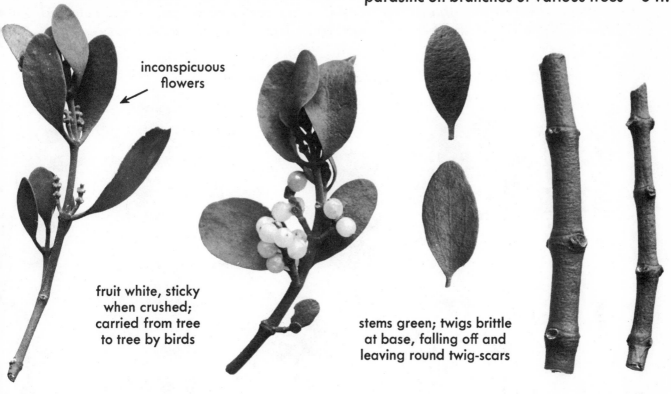

inconspicuous
flowers

fruit white, sticky
when crushed;
carried from tree
to tree by birds

stems green; twigs brittle
at base, falling off and
leaving round twig-scars

JAPANESE KNOTWEED—*Polygonum cuspidatum* Perfect or Polygamous   Aug.
Asia; commonly escaped elsewhere   6-8 ft.   Alternate

This is not a true woody shrub, but a perennial plant.
It is included here because it is often mistaken for Bam-
boo with its hollow, woody-appearing stems. These die
back to the ground during the winter, new ones coming
up each spring.

BARBERRY—*Berberis* Perfect Alternate

American Barberry	*B. canadensis*	May-June	s. Pa. & W. Va., s. to Ga. & s.e. Mo.	3 (occ. 5)
Common Barberry	*B. vulgaris*	May-June	Europe; planted here, commonly escaped	7-8
Japanese Barberry	*B. thunbergii*	April-May	Japan; planted here, commonly escaped	7-8

All Barberry flowers are yellow.

red

red

Common Barberry
(American B. similar,
but with fewer flowers
in each raceme)

Common
Barberry

red

American Barberry

Japanese Barberry

this leaf a
juvenile
form

Common
Barberry

American Barberry

MP
18

184

The inner bark of all Barberries is yellow.

Twigs and young stems are reddish brown in winter; older stems are gray (top row).

thorns curving, mostly triple
(except sometimes single on small twigs)

thorns straight and slender, mostly single
(except occasionally triple on largest stems)

American Barberry

Japanese Barberry

Twigs and all older stems are uniformly gray in winter (bottom row).
Thorns are mostly triple (or more),
except sometimes single on small twigs.

Common Barberry

SWEET SHRUB—*Calycanthus floridus* Perfect April-July Opposite
(Carolina Allspice; Hairy Strawberry Shrub)   Pa. & W. Va. to Fla. & Miss. 9 ft.

Note: The Smooth Strawberry Shrub—*C. fertilis*—is found in much the same range, the principal differences being that the flowers are less fragrant and the leaves relatively hairless beneath.

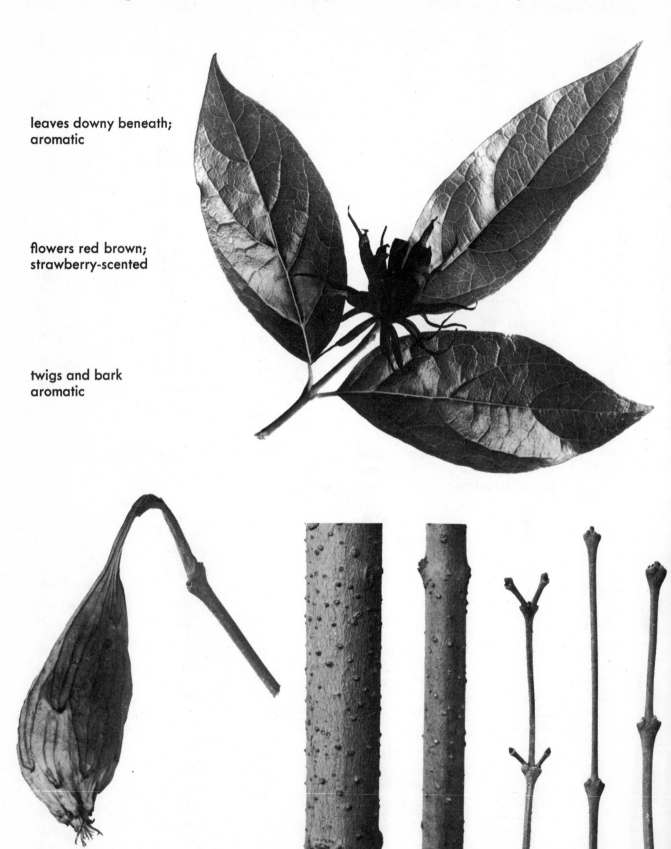

leaves downy beneath; aromatic

flowers red brown; strawberry-scented

twigs and bark aromatic

MP
20

SPICE-BUSH—*Lindera benzoin*   Dioecious or Polygamous   March-April   Alternate
(Spiceweed; Benjamin Bush)   Me. to Ont. & Mich., s. to Fla. & Tex.   15 ft.

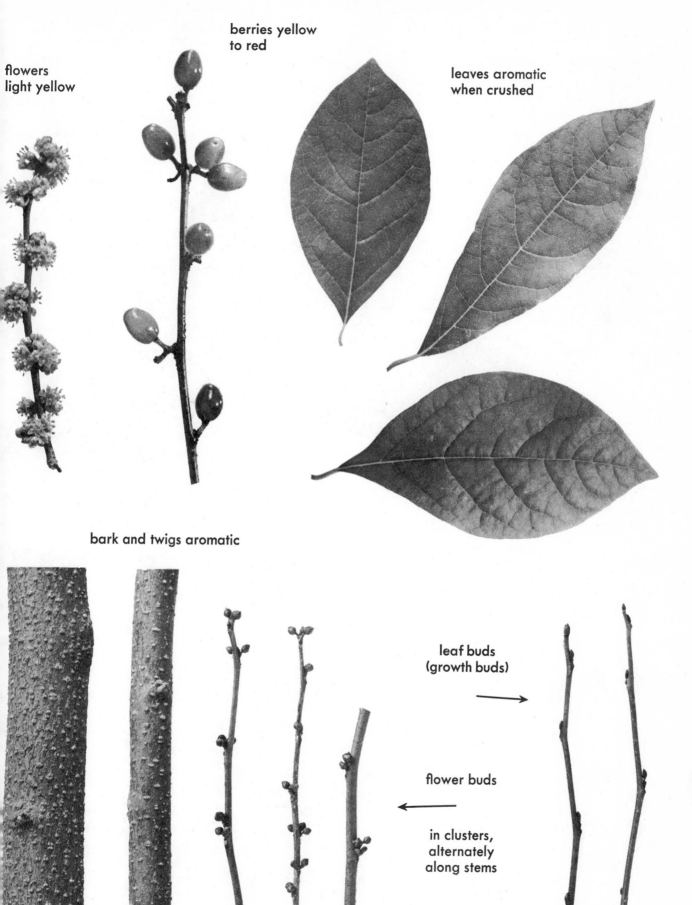

berries yellow
to red

flowers
light yellow

leaves aromatic
when crushed

bark and twigs aromatic

leaf buds
(growth buds)

flower buds

in clusters,
alternately
along stems

MP
21

MOCK-ORANGE—*Philadelphus* Perfect Opposite

Large-flowered Mock-orange	*P. grandiflorus*	May-June	Va. & Tenn., s. to Fla. & Miss.; commonly planted in n. states	9
Scentless Mock-orange	*P. inodorus*	May-June	Va. & Tenn., s. to Fla. & Miss.; commonly planted in n. states	9
Sweet Syringa (Mock-orange)	*P. coronarius*	May-June	s. Europe; commonly planted here	9

note: usually three blossoms

note: usually more than three blossoms

very fragrant

Large-flowered Mock-orange
(Scentless M. similar)

Sweet Syringa

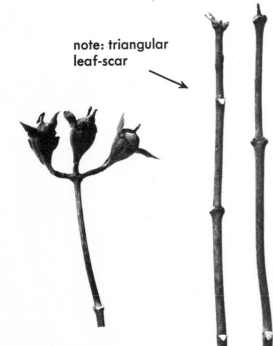

note: triangular leaf-scar

Scentless Mock-orange
(Large-flowered M. similar)

twigs and bark typical of Mock-oranges

Sweet Syring

mostly without teeth

mostly with teeth

Scentless Mock-orange

Sweet Syringa
(Large-flowered M. similar)

VIRGINIA-WILLOW — *Itea virginica*   Perfect   May-June   Alternate
(Tassle-white)
N. J., e. Pa. & Va. to Okla.,
s. to Fla. & Tex.   4 (occ. 9) ft.

young stems and twigs
green, tinged with red

Leaf-stalk often re-
mains on twig when
main part of leaf
falls off.

FALSE SPIREA—*Sorbaria sorbifolia* Perfect   June-July   Alternate
(Mountain Ash Spirea)
n. Asia; planted here & escaped   6 ft.

MP
24

SHRUB YELLOW-ROOT—*Xanthorhiza simplicissima*

Polygamous April-May Alternate
N. Y. to W. Va., s. to Fla. & Ala.   2 ft.

flowers
brownish purple

roots and inner bark
of stems yellow

MP
25

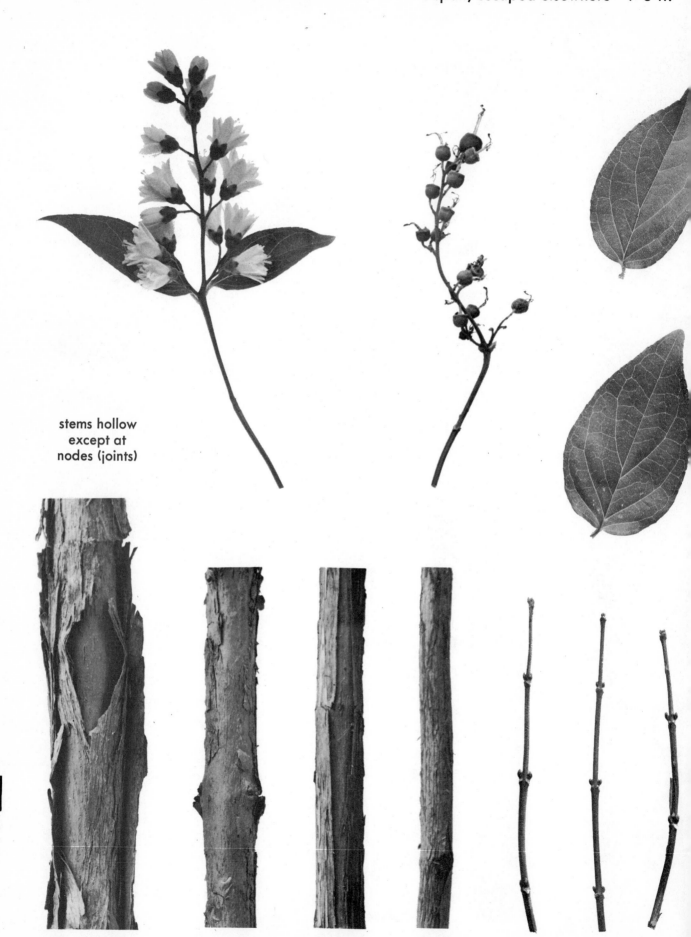

stems hollow
except at
nodes (joints)

MP
26

192

NINEBARK—*Physocarpus opulifolius* Perfect June Alternate
Que. to Minn. & Colo., s. to S. C. & Ark.   9 ft.

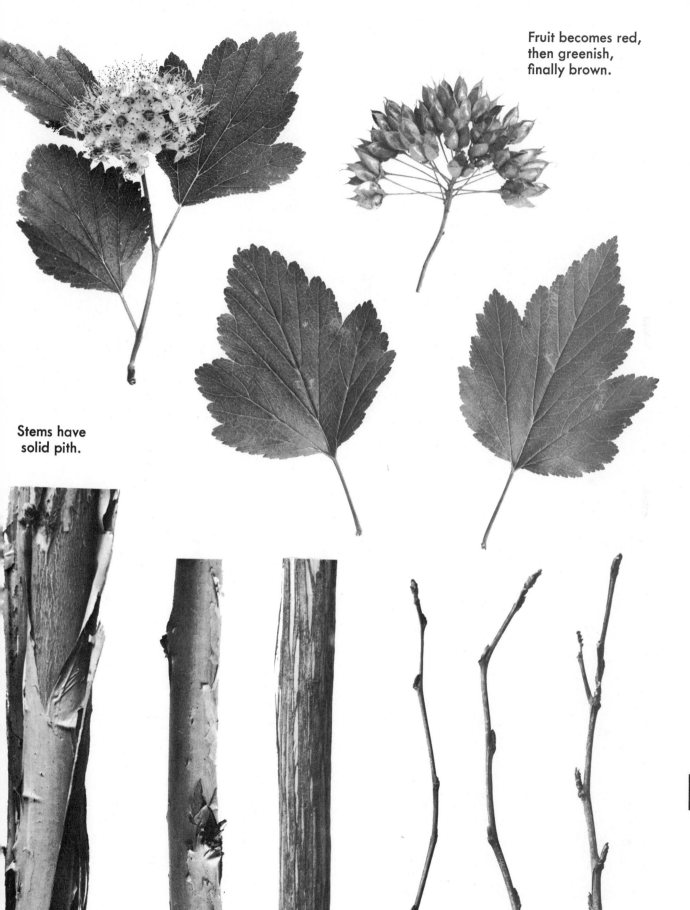

Fruit becomes red,
then greenish,
finally brown.

Stems have
solid pith.

HYDRANGEA — *Hydrangea*   Flowers of two kinds: Perfect and Sterile   Opposite

Peegee Hydrangea   *H. paniculata grandiflora*   Mostly sterile flowers   Aug.-Sept.
Japan & China 15-20 (occ. 30) ft

Wild Hydrangea   *H. arborescens*   Mostly perfect flowers with a few marginal sterile ones   June-
(Seven-bark)   N. Y. to s. Ia., s. to Fla. & La. 3-4 (occ. 10)

Note: A variety of Wild Hydrangea known as Hills of Snow — *H. a. grandiflora* — has mostly sterile flowers.

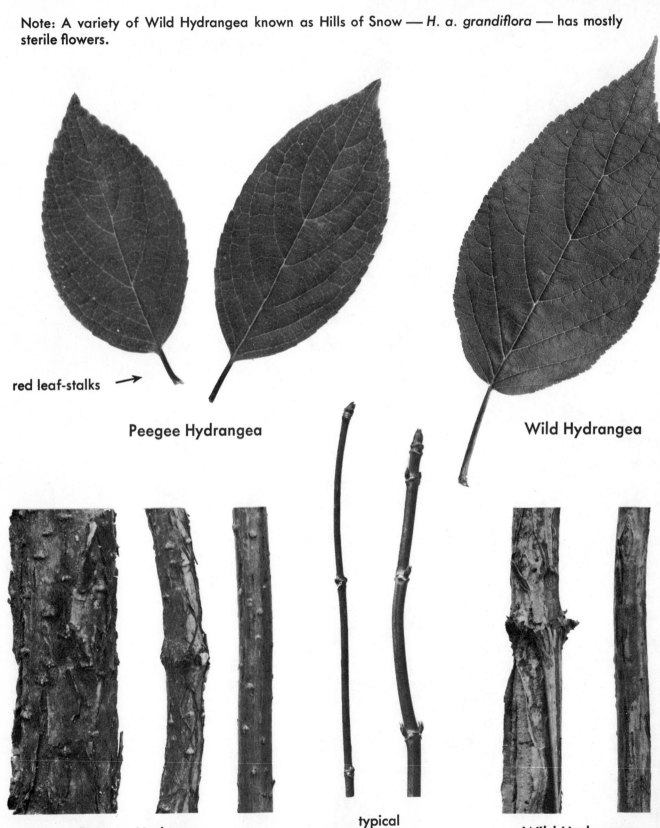

red leaf-stalks →

Peegee Hydrangea

Wild Hydrangea

Peegee Hydrangea

typical
Hydrangea twigs

Wild Hydrangea

pyramidal flower clusters;
flowers mostly sterile

As most of the flowers are
sterile, there is little actual
fruit, but the flower remnants
last a considerable time.

Peegee Hydrangea

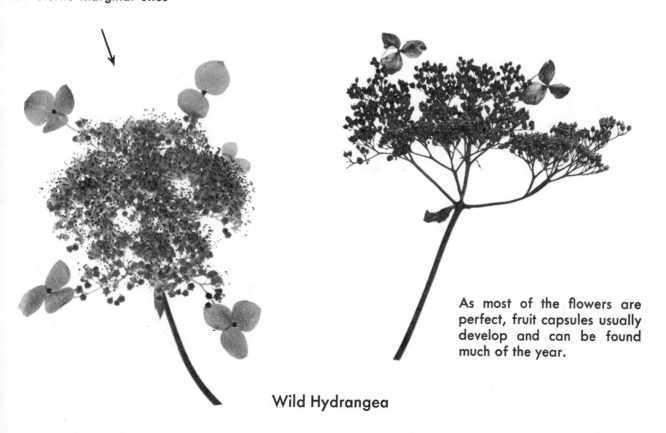

flat flower cluster; flowers
mostly perfect, with only a
few sterile marginal ones

As most of the flowers are
perfect, fruit capsules usually
develop and can be found
much of the year.

Wild Hydrangea

MP
29

CURRANT and GOOSEBERRY—*Ribes* Perfect Alternate
(Dioecious in some foreign species not included here)

I. GOOSEBERRIES: flowers (and fruit) few (one to five) from any one point on stems

European Gooseberry	*R. grossularia*	May-June	Europe, n. Africa & Asia; escaped here 3½ f
Prickly Gooseberry	*R. cynosbati*	May-June	N. B. to Man., s. to N. C., n. Ala. & Mo. 6 f
Smooth Gooseberry	*R. hirtellum*	April-July	Lab. to Man., s. to N. C. & Tenn. 3 f
(Low Wild Gooseberry)			

Note: Considerable confusion exists concerning the actual differences between the Smooth Gooseberry and the Round-leaved Gooseberry — *R. rotundifolium* — and until further study indicates two clear-cut species, it would seem pardonable to consider them one species, with perhaps some local but unspectacular variations. The two are treated as one here, and the range given is for both types.

II. CURRANTS: flowers (and fruit) in racemes of five or more individual flowers (or fruit)

American Black Currant	*R. americanum*	April-June	N. S. to Alb., s. to Va., Mo. & west 6 f
(Wild Black Currant)			
Bristly (Black) Currant	*R. lacustre*	May-Aug.	Nfd. to Alaska, s. to n. N. E., w. Mass., N. Y., Pa., n. Ohio, Mich., Wis., Minn. & west 3 f
European Black Currant	*R. nigrum*	April-June	Europe & Asia; escaped here 6 f
Garden Red Currant	*R. sativum*	April-June	Europe; escaped here 5 f
Skunk Currant	*R. glandulosum*	May-Aug.	Lab. to B. C., s. to s. N. E., N. Y., Pa., W. Va., mt of N. C., Ohio, Mich., Wis. & Minn.; stems reclinin or creeping, with ascending branches 2 f

Note: Most parts of this plant have skunk-like odor when crushed, especially the twigs.

Swamp Red Currant	*R. triste*	April-June	Lab. to Alaska, s. to s. N. E., N. Y., Pa., W. Va., Ill la. & west; arching or straggling stems 2½ f

I. GOOSEBERRIES: flowers few (one to five) from any one point on stem

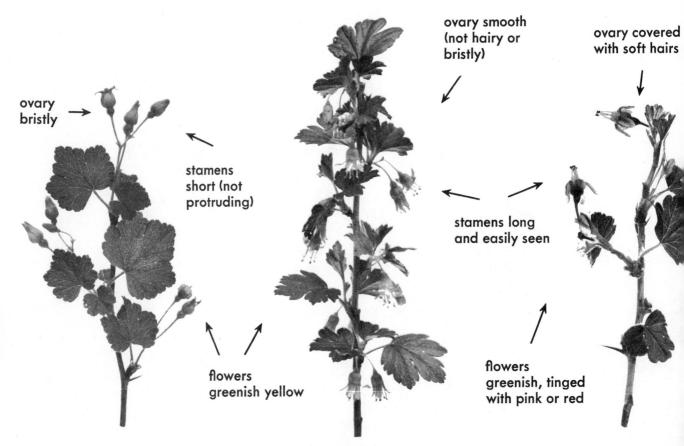

ovary bristly

stamens short (not protruding)

ovary smooth (not hairy or bristly)

ovary covered with soft hairs

stamens long and easily seen

flowers greenish yellow

flowers greenish, tinged with pink or red

Prickly Gooseberry   Smooth Gooseberry   European Goosebe

II. CURRANTS: flowers in racemes (of five or more individual flowers)

A. flowers flat, saucer-shaped

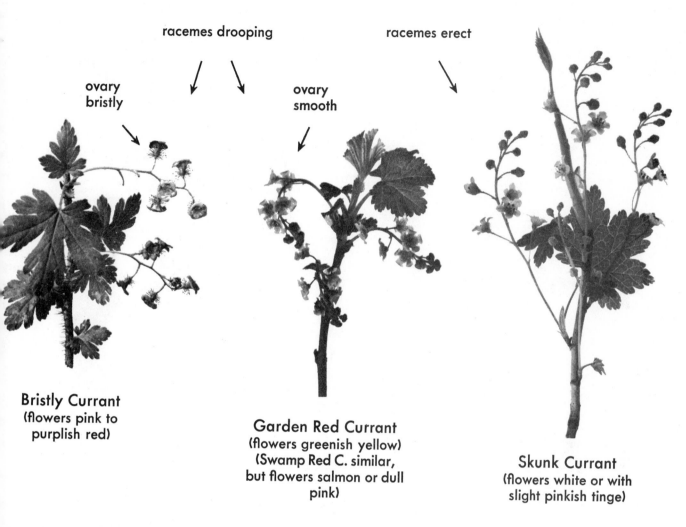

racemes drooping

racemes erect

ovary bristly

ovary smooth

Bristly Currant
(flowers pink to purplish red)

Garden Red Currant
(flowers greenish yellow)
(Swamp Red C. similar, but flowers salmon or dull pink)

Skunk Currant
(flowers white or with slight pinkish tinge)

B. flowers bell-shaped

American Black Currant
(flowers pale yellow)

European Black Currant
(flowers greenish, tinged with red or pink)

MP 31

I. GOOSEBERRIES

dull red

dull red

Smooth Gooseberry

Prickly Gooseberry

red, yellow or green

European Gooseberry

II. CURRANTS

bristly berries

black

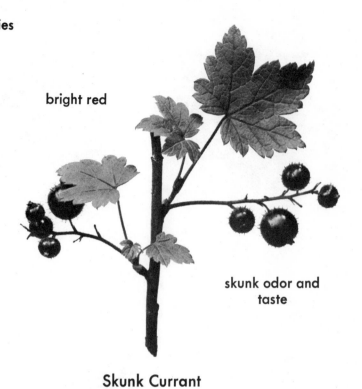

bright red

skunk odor and
taste

Skunk Currant

Bristly Currant

smooth berries

shiny black

bright red

American Black Currant
(European Black C. similar)

Garden Red Currant
(Swamp Red C. similar,
except slightly smaller,
harder and less juicy)

MP
33

leaves
velvety
underneath

Prickly Gooseberry

Smooth Gooseberry

MP
34

teeth
generally
more rounded
than those of
Smooth Gooseberry

European Gooseberry

Bristly Currant

hairy
leaf-stalk

Skunk Currant

Swamp Red Currant

reddish
leaf-stalk

resin-
dotted both sides
(more clearly so
beneath; use lens)

American Black Currant

Garden Red Currant

resin-
dotted
beneath only

European Black Currant

MP
35

All Gooseberries are armed (have some thorns or bristles).

Prickly Gooseberry

Smooth Gooseberry

The Bristly Currant is the only armed Currant. T small stems and twigs are golden brown a densely bristly.

European Gooseberry

Bristly Currant

All Currants are unarmed (no thorns or bristles) with the exception of the Bristly Currant shown on the opposite page.

American Black Currant

Garden Red Currant

Stems and twigs have
skunk odor when broken.

brown buds ↙   ↓   red buds ↙

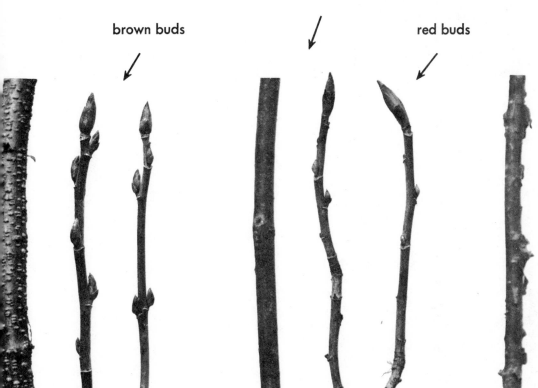

ِuropean Black Currant   **Skunk Currant**   **Swamp Red Currant**

WITCH-HAZEL—*Hamamelis virginiana* Perfect Sept.-Nov. Alternate
Que. to Minn., s. to Ga. & Ark.   15 ft.

pale yellow flowers

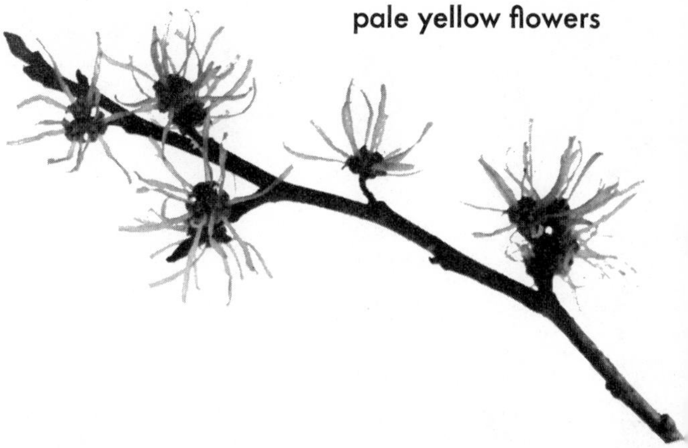

Fruit capsule bursts open when ripe, expelling seeds for wide distances (up to 50 ft.).

naked buds
(without scales)

MP
38

204

CHOKEBERRY—*Aronia*   Perfect   April-May   Alternate

Black Chokeberry	*Aronia melanocarpa*	N. S. to Ont., s. to S. C., Tenn. & Ia.	3-5 ft.
Red Chokeberry	*A. arbutifolia*	s. N. E., Ont. & Ohio, s. to Fla. & Tex.	10 ft.

Note: Purple Chokeberry — *A. prunifolia* — N. S. to Ont. & Mich., s. to Va. & Ill., appears to be an intermediate species, with characteristics between the other two Chokeberries. The fruit is dark red or purple; the height is variable, but up to 10 ft.

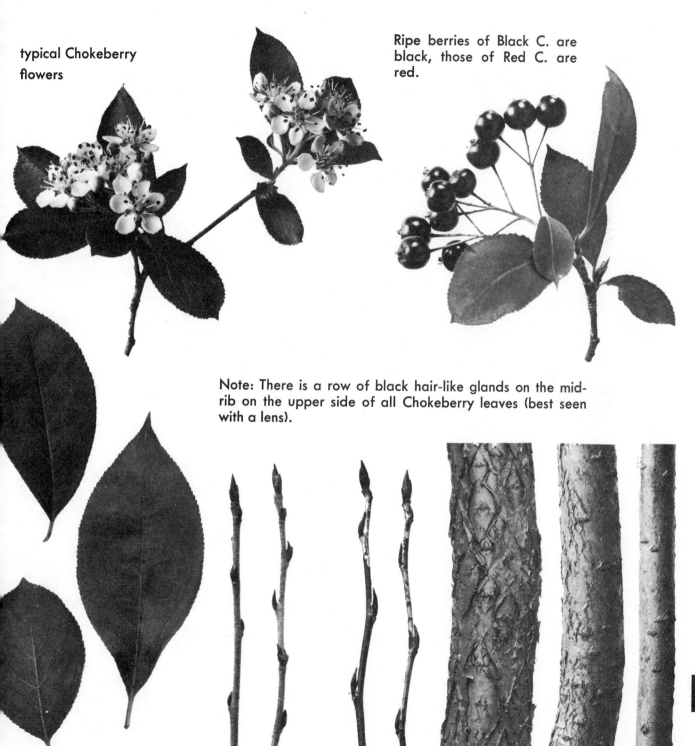

typical Chokeberry flowers

Ripe berries of Black C. are black, those of Red C. are red.

Note: There is a row of black hair-like glands on the mid-rib on the upper side of all Chokeberry leaves (best seen with a lens).

typical Chokeberry leaves
Black C.: smooth beneath
Red C.: fuzzy beneath

Red Chokeberry (fuzzy twigs)

Black Chokeberry (smooth twigs, often with a silvery coating)

typical Chokeberry barks

SPIREA—*Spiraea* Perfect Alternate

Corymbed Spirea *Spiraea corymbosa* (Birch-leaved Meadowsweet)	June-July	s. Pa. to Ga. & Ky.	3 ft.
Hardhack *S. tomentosa* (Steeplebush)	July-Sept.	N. S. to Ont. & Minn., s. to Ga. & Ark.	3-4 ft.
Meadowsweet *S. latifolia* (Broad-leaved Meadowsweet)	June-Aug.	Nfd. to Mich., s. to N. C.	4-5 ft.

Note: There are a large number of foreign Spireas much used by nurseries in this country, but few, if any, have become naturalized.

pink

white

flat flower cluster

pale beneath;
often rusty
and woolly

green
beneath

usually n
teeth bel
the midd

Hardhack

Meadowsweet

Corymbed Spirea

MP
40

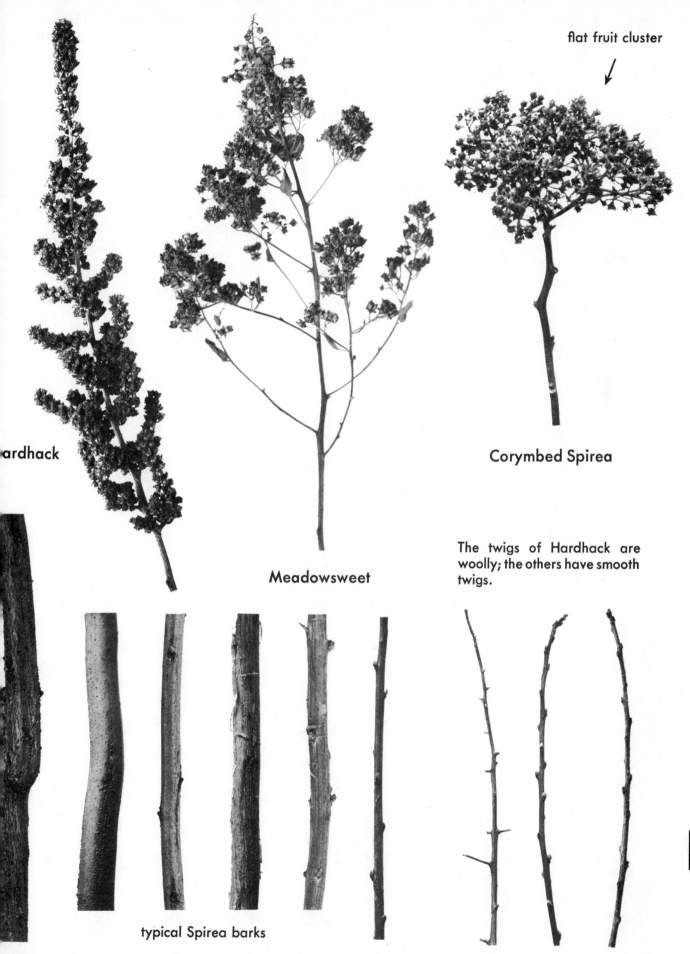

flat fruit cluster

ardhack

Corymbed Spirea

Meadowsweet

The twigs of Hardhack are woolly; the others have smooth twigs.

typical Spirea barks

typical Spirea twigs

Older barks are gray; young shoots are brown, sometimes smooth and shiny, at other times somewhat shaggy.

MP 41

FIRETHORN—*Pyracantha coccinea* *lalandii* Perfect May-June Alternate
Evergreen s. Europe & w. Asia; naturalized Pa. & south,
hardy farther north 8 (occ. 20) ft.

orange-red berries

usually very
thorny

MP
42

HAWTHORN — *Crataegus*   Perfect   mostly April-June   shrubs or trees   Alternate

There are hundreds of species in the genus, at one time over one thousand being named in North America alone. However, many of these are no longer considered to be separate species, but rather hybrids, varieties or forms.

Note variation in leaf shapes.

usually very thorny

round little red buds, usually at the ends of small spurs

MP 43

This is a large and confusing genus; there is considerable difference of opinion concerning the various species.

Fruit turns from pink
or red to purple or black.

MP
44

210

NEW JERSEY TEA—*Ceanothus americanus* Perfect   June-Sept.   Alternate
Canada, s. to Fla. & Tex. 3 ft.

fruit at first covered
by black case, opening
to expose three white nutlets

JETBEAD—*Rhodotypos scandens*   Perfect   May-June   Opposite
Japan, China; escaped elsewhere 6 ft.

black berries

MP
45

BRAMBLE (Raspberries, Blackberries, Dewberries) — *Rubus*   Perfect   Alternate

I. RASPBERRIES

American Red Raspberry	*Rubus idaeus strigosus*	May-June	Nfd. to B.C., s. to Va., Tenn. & Wyo.; stems erect (not arching or rooting at tips) 6 ft.

(This is considered a variety of the European Red Raspberry—*R. idaeus*.)

Black Raspberry *R. occidentalis* (Black-cap; Thimbleberry)		May-June	N.B. to Minn., s. to Ga., Tenn. & west; stems (canes) arching up to 10 ft. long and rooting at tips
Purple Flowering Raspberry *R. odoratus* (Thimbleberry)		June-Aug.	N.S. to Mich., s. to Tenn. & Ga.; stems erect 6-8 ft.
Wineberry *R. phoenicolasius*		June-July	e. Asia; escaped in e. U.S.; arching stems to 10 ft. long and rooting at tips

II. BLACKBERRIES and DEWBERRIES

Highbush Blackberry *R. allegheniensis* (Allegheny Blackberry)		May-July	N.S. to Minn., s. to N.C. & Ark. 10 ft.
Smooth Blackberry *R. canadensis* (Thornless B.; Canada B.)		June-July	Nfd. to Ont. & Minn., s. to Ga., Tenn. & Ill. 10 ft.
Common Dewberry *R. flagellaris*		May-June	Me. to Minn., s. to Va. & Mo. Prostrate
Swamp Dewberry *R. hispidus*		June-July	N.S. to Mich., s. to Ga. & Ill.; semi-evergreen Prostrate

This genus, especially the subgenus Blackberry, is a very confusing one, as it is undergoing rapid evolutionary changes and therefore the species are not stable. It is thought that this variability is a result of extensive clearing of land and other environmental changes beginning with the arrival of the white man. However, as with other difficult genera, the main types are sufficiently clear-cut for practical purposes.

> Note: The fruit of Raspberries separates easily from what is called the receptacle. A definite hole at the base of the fruit marks the point of separation, and a corresponding projection or knob is left on the stem (see Black Raspberry fruit, MP 50). This distinguishes all Raspberries from the Blackberries and Dewberries, which do not separate from the receptacle in this manner.

MP
46

1. RASPBERRIES

(see also next page)

leaves green on both sides;
leaf-stalk sticky, hairy

Purple Flowering Raspberry

leaves white beneath;
leaf-stalk sticky, hairy

Wineberry

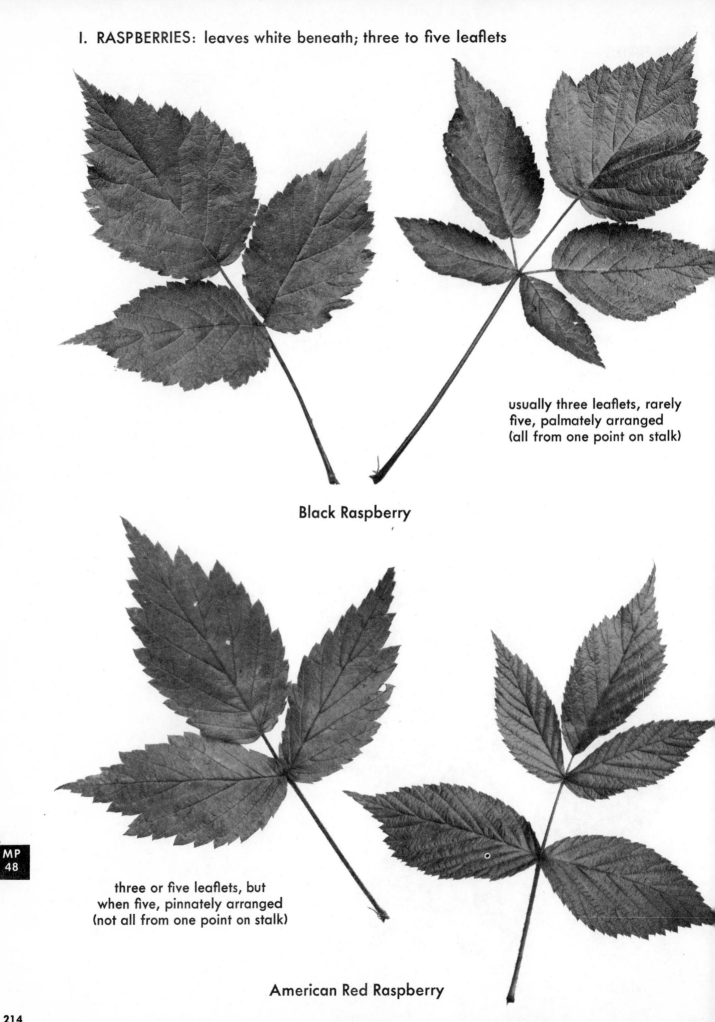

I. RASPBERRIES: leaves white beneath; three to five leaflets

usually three leaflets, rarely
five, palmately arranged
(all from one point on stalk)

Black Raspberry

MP
48

three or five leaflets, but
when five, pinnately arranged
(not all from one point on stalk)

American Red Raspberry

II. BLACKBERRIES and DEWBERRIES:

leaves green beneath; three or five leaflets; when five, always palmately arranged

densely covered with woolly hairs on underside, velvety to touch

Highbush Blackberry
(Smooth B. similar but not velvety beneath)

generally very glossy; semi-evergreen, but often with reddish cast, especially in fall and winter; usually three, rarely five leaflets

three to five leaflets

MP 49

Swamp Dewberry

Common Dewberry

I. RASPBERRIES

Black Raspberry
(Red R. similar)

magenta, turni
lavender pink

**Purple Flowering
Raspberry**

Note white projec-
tion left after fruit
is removed (typical
of Raspberries).

Black Raspberries
are red before ripening,
but black when ripe.

Red Raspberries
are red when ripe.

Black Raspberry
(Red R. similar)

red; rath
tastele

**Purple Flowering
Raspberry**

stems and
parts of flowers
(and fruit) densely
red-hairy; hairs
gland-tipped and sticky

red; rath
tasteles

Wineberry

MP
50

II. BLACKBERRIES and DEWBERRIES

Swamp Dewberry
(Common D. similar, but
petals usually larger)

Highbush Blackberry
(Smooth B. similar, but
flower stems hairless)

fruit sour;
black when
ripe

hairy
stems

Swamp Dewberry

fruit sweet;
black when ripe

fruit sweet;
black when ripe

Common Dewberry

Highbush Blackberry
(Smooth B. similar)

MP
51

I. RASPBERRIES

arching; rooting at tips; heavy white bloom over red stems

erect or arching; not rooting at tips; sli (occ. heavy) bloom; brown or red stems

Black Raspberry

American Red Raspberry

erect; not rooting at tips

arching; rooting at tips; densely red-hairy, with some thorns

Purple Flowering Raspberry

Wineberry

MP
52

II. BLACKBERRIES and DEWBERRIES: stems green or red without bloom

Blackberries: large; upright or arching

Dewberries: prostrate, or scrambling over low objects

stems round, or often grooved

unarmed, or with a
few weak prickles

semi-evergreen;
usually very bristly slender stems

deciduous;
not bristly, but with a few stiff prickles

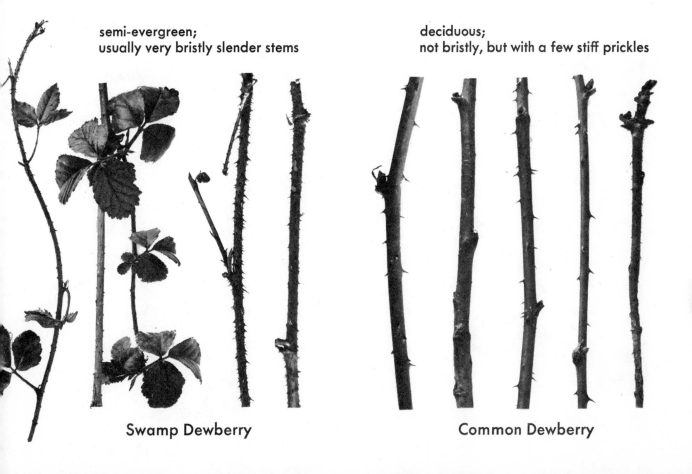

Smooth Blackberry

Highbush Blackberry (stout thorns usually present)

Swamp Dewberry

Common Dewberry

MP
53

CHERRY and PLUM—*Prunus*   Perfect   Alternate

Beach Plum *Prunus maritima*	April-June	Me. to Va. near coast	6-8 f
Choke Cherry *P. virginiana*	May	Nfd. to Sask. & S. Dak., s. to N. C., Kans. & west	30 f
Sand Cherry *P. pumila*	April-early June	w. N. Y. to Wis., but in var. Me. & Mass., w. to Minn. & Ill.	3 (occ. 8) f
Wild Plum *P. americana* (Wild Yellow or Red Plum)	April-early June	Vt. & Ont. to Minn., s. to Fla. & N. Mex.	25 f

flowers (and fruit) in terminal racemes

flowers (and fruit) along stems; (few or many, but not in terminal racemes)

Choke Cherry

Wild Plum

(Beach P. and Sand Cherry similar)

red, turning blue black

Choke Cherry

Sand Cherry

purple or black

red or purple

Beach Plum

red, rarely yellow

Wild Plum

Note: Black Cherry, a tree, is sometimes confused with Choke Cherry, but the leaves of Black Cherry have rounded teeth, are glossier and usually narrower than those of Choke Cherry.

Choke Cherry

teeth usually few or missing along bottom third of leaf

Sand Cherry

Wild Plum

Beach Plum

Twigs and bark have an unpleasant smell when broken.

tent-caterpillar egg mass, often found on Cherries and Plums

usually somew[hat] thorny (see a[lso] Thorn Key, p.

Choke Cherry   Sand Cherry   Beach Plum   Wild Plum

bark usually not scaly until larger than this

Choke Cherry   Beach Plum
(Sand Cherry similar)   Wild Plum

ROSE—*Rosa* Perfect Alternate

Multiflora Rose	*R. multiflora*	June-July	e. Asia; widely planted, escaped; arching shrub to 8 ft., rambling if support is available to 10-12 ft.
Pasture Rose (Carolina Rose)	*R. carolina*	June-July	N.S. to Minn., s. to Fla. & Tex.; small shrub 3 ft.
Prairie Rose	*R. setigera*	June-July	c. N.Y. to Mich., Ill. & Kans., s. to Fla. & Tex.; arching shrub to 6 ft., clambering if support is available, but not to the extent of the Multiflora R.
Rugosa Rose	*R. rugosa*	May-Aug.	e. Asia; widely planted, escaped 6 ft.
Smooth Rose (Meadow Rose)	*R. blanda*	May-July	Nfd. to Man., s. to n. & w. N.E., N.Y. & Pa., w. to Mo. 6 ft.
Swamp Rose	*R. palustris*	June-Aug.	N.S. to Minn., s. to Fla. & Ark. 8 ft.
Sweet-Brier (Eglantine)	*R. eglanteria*	May-July	Europe; naturalized here 10 ft.
Virginia Rose	*R. virginiana*	June-July	Nfd. to s. Ont., N.Y. & N.E., s. to N.C., Ala., Tenn. & Mo., but apparently rare or absent in states w. & n. of Pa. 6 ft.

The Roses comprise a beautiful but confusing group of plants. There have been many differences of opinion concerning the various species, no two experts ever totally agreeing on the relative importance of apparently conflicting factors. In addition to man-made confusion, the Roses themselves contribute their own peculiar ways to complicate the situation. As a group they have distinctive features that make them known at once as Roses, but the individuals, within many species, appear to be deliberately perverse, and sports, variations and hybrids are common, making for instability within species. It is important to bear this in mind when looking for acceptable features on which to base species identification.

There are certain characteristics, nevertheless, that can be depended upon to remain relatively stable and which can be used to divide the Roses into distinct groups. It is best to attempt species identification beginning at flowering time and continuing through the fruiting period, although on better acquaintance it is often possible to identify many Roses at any season. As with a few other genera included in this book, main groupings are the primary objectives, and representative species of only the most common types are shown.

Always look at all available details; a combination of details, not just one, is essential for the proper identification of Rose species.

MP
57

Note: The pistils (female part of flower) of Multiflora Rose and Prairie Rose are long, forming a protruding column from the center of the flower.

pink to white

white

Note: Both of these are Rambler Roses; the others are erect or arching but do not clamber.

Prairie Rose

Multiflora Rose

note pistils not extended; centers of flowers flat

magen

pink to white

Smooth Rose

(Pasture R., Swamp R., Sweet-Brier and Virginia R. similar)

Rugosa Rose

Note: Roses bear a red fruit called a "hip." This is distinctive of the genus and in some cases helpful in distinguishing the species. The Rugosa and Smooth Roses retain their sepals (leaf-like appendages at the end of the hip); the others lose their sepals by the time the fruit is ripe, or earlier.

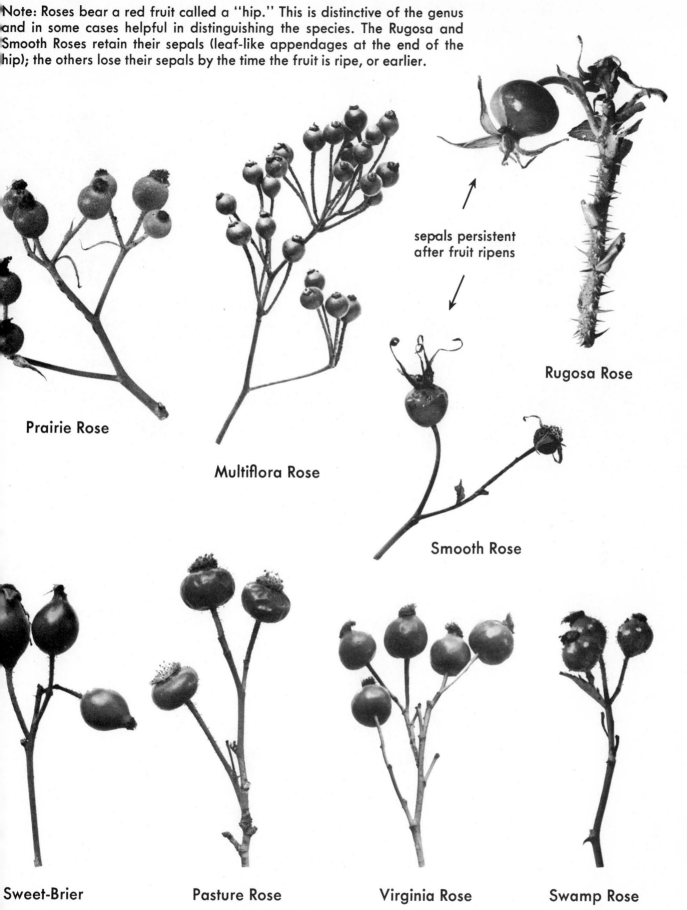

sepals persistent after fruit ripens

Rugosa Rose

Prairie Rose

Multiflora Rose

Smooth Rose

Sweet-Brier

Pasture Rose

Virginia Rose

Swamp Rose

MP 59

Note: The fruit of these four Roses vary considerably and should not be used to determine the species. They are shown here to indicate characteristics possible for any one.

Note: Rose leaves bear stipules (leafy growths attached to the leaf-stalk) characteristic of the genus and in some cases typical of a particular species (see Multiflora and Swamp Roses).

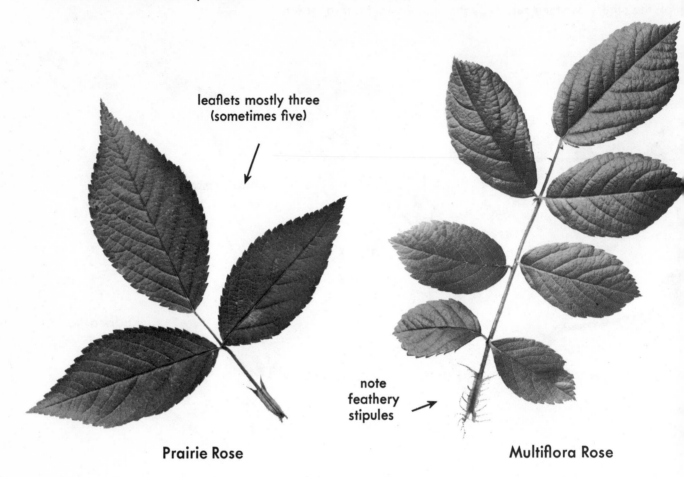

leaflets mostly three (sometimes five)

note feathery stipules

Prairie Rose

Multiflora Rose

**Pasture Rose
(Virginia Rose similar)**

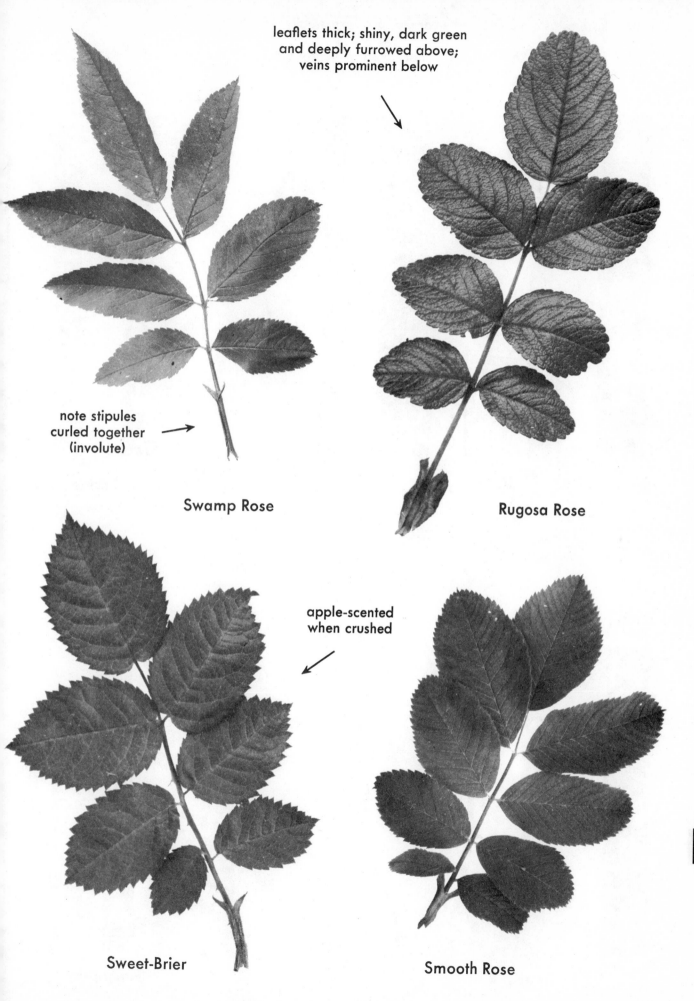

leaflets thick; shiny, dark green
and deeply furrowed above;
veins prominent below

note stipules
curled together
(involute)

Swamp Rose

Rugosa Rose

apple-scented
when crushed

Sweet-Brier

Smooth Rose

MP
61

227

Note: Rose twigs are usually green or red; buds are usually red.

Thorns of Multiflora Rose may be light-colored or nearly black; they are usually heavy, but there is a thornless variety.

Multiflora Rose

Prairie Rose

Swamp Rose

Sweet-Brier

rough, blackish bark

few if any thorns; weak bristles
on lower part of large stems only

Rugosa Rose

Smooth Rose

thorns mostly curved

thorns mostly straight, needle-like

Virginia Rose

Pasture Rose

MP
63

CINQUEFOIL—*Potentilla* Perfect Alternate

Shrubby Cinquefoil	*P. fruticosa*	May-Aug.	N. Hemisphere; in N. Amer., Arc tic, s. to N. J., Pa., Ill., Ia., S. Dak. N. Mex. & west 2-3 ft
Three-toothed Cinquefoil	*P. tridentata*	May-Sept.	Arctic, s. to N.E. & N.Y., w. to N Dak., s. in mts. to Ga.; subshrub Prostrate

yellow flowers

white
flowers

MP
64

Shrubby Cinquefoil

Three-toothed Cinquefoil

DYER'S GREENWEED—*Genista tinctoria*
(Woad Waxen; Whin)

Perfect June-Aug. Alternate
Europe & Asia; naturalized e. states;
semi-evergreen   3 ft.

yellow
flowers

twigs and small stems green

SCOTCH BROOM—*Cytisus scoparius*

Perfect May-June Alternate
Europe; naturalized e. states   6 ft.

yellow
flowers

twigs and small stems green

Well-developed leaves
have three leaflets, but
the upper leaves are
often reduced to one,
and sometimes few
leaves develop at all, the
stems being bare even
in summer.

MP
65

FALSE INDIGO – *Amorpha* Perfect   Alternate

Indigo Bush   *A. fruticosa*   May-June   Mich. to Minn., s. to Fla. & La.; planted and es
(False Indigo)   caped in n.e. states   12 ft
Lead-plant   *A. canescens*   June-July   Mich. to Sask., s. to Ind., Tex. & N. Mex.   3 ft
(Shoestring)

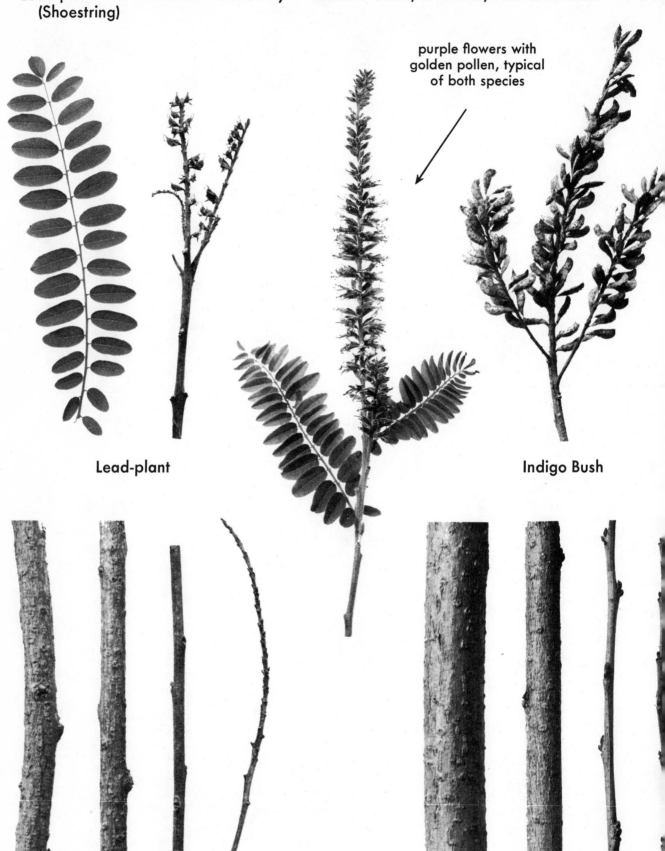

purple flowers with
golden pollen, typical
of both species

Lead-plant

Indigo Bush

Lead-plant

Indigo Bush

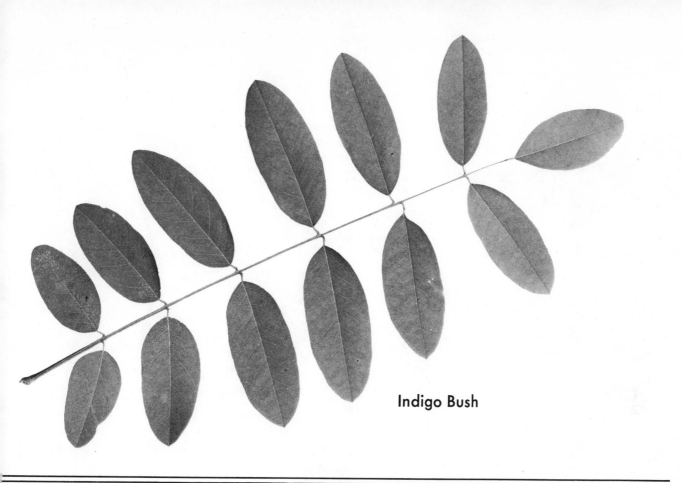

Indigo Bush

GORSE—*Ulex europaeus*   Perfect   March-June   Alternate
Europe; naturalized Atlantic states and
West Coast   3-(rarely) 6 ft.

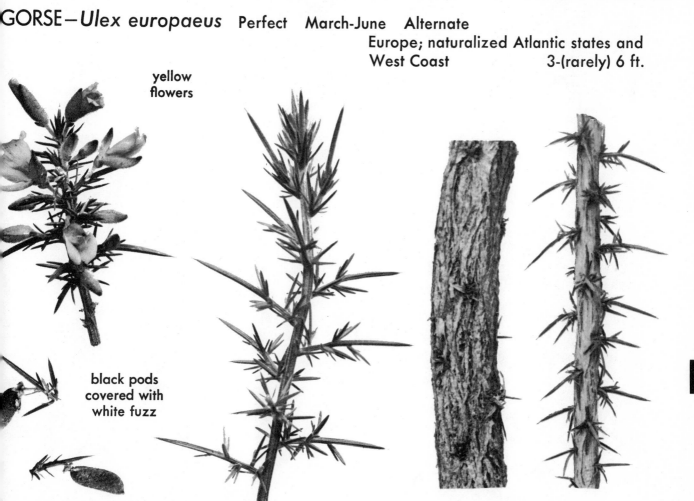

yellow
flowers

black pods
covered with
white fuzz

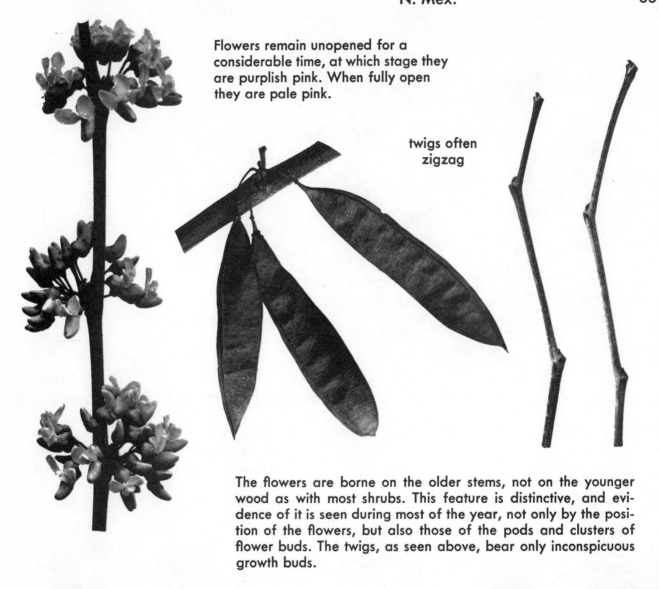

Flowers remain unopened for a considerable time, at which stage they are purplish pink. When fully open they are pale pink.

twigs often zigzag

The flowers are borne on the older stems, not on the younger wood as with most shrubs. This feature is distinctive, and evidence of it is seen during most of the year, not only by the position of the flowers, but also those of the pods and clusters of flower buds. The twigs, as seen above, bear only inconspicuous growth buds.

flower buds

234

BRISTLY LOCUST—*Robinia hispida* Perfect May-June Alternate

Va. to Ky., Ga. & Ala.; planted & escaped northward   3-4 ft. (occ. more)

pink flowers

MP
69

PRICKLY ASH—*Zanthoxylum americanum*
(Toothache-tree)

Dioecious or Polygamous   April-May   Alternat
Que. to N. Dak., s. to Ga. & Okla. 25 f

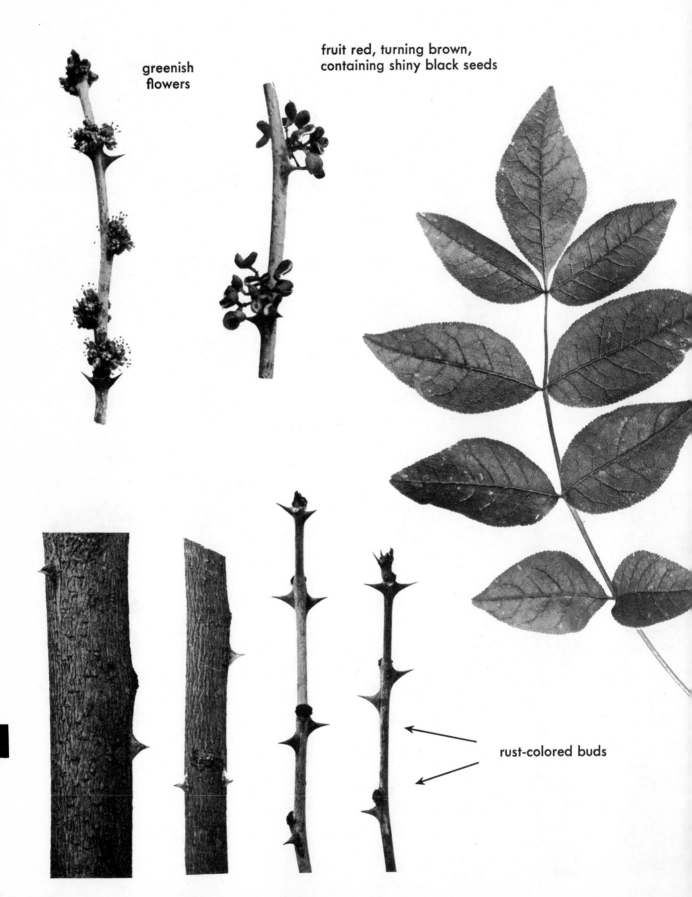

greenish flowers

fruit red, turning brown, containing shiny black seeds

rust-colored buds

HOP-TREE—*Ptelea trifoliata*   Polygamous   May-June   Alternate
s.w. Que. & N.Y. to s. Ont. & Mich., s.
to Fla. & Tex.; escaped elsewhere 25 ft.

flowers
greenish white

SMOKE-TREE—*Cotinus*   Dioecious or Polygamous   Alternate

American Smoke-tree   *C. americanus*   Dioecious   May-July   Tenn. & Mo. to Tex. & Al
  (Chittam-wood)   35
Smoke-tree   *C. coggygria*   Polygamous   May-July   s. Europe, c. China & Hi
                                                             alayas; planted here & o
                                                             escaped   15

tiny individual flower
stems covered with
pinkish hairs

flowers of
both species
yellowish

X

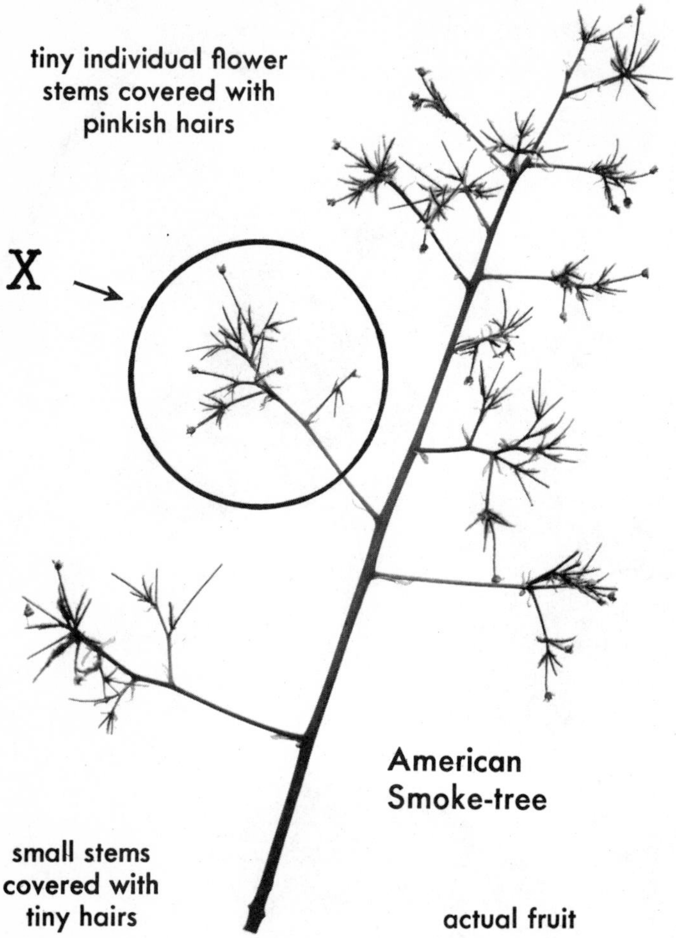

American
Smoke-tree

Smoke-tree

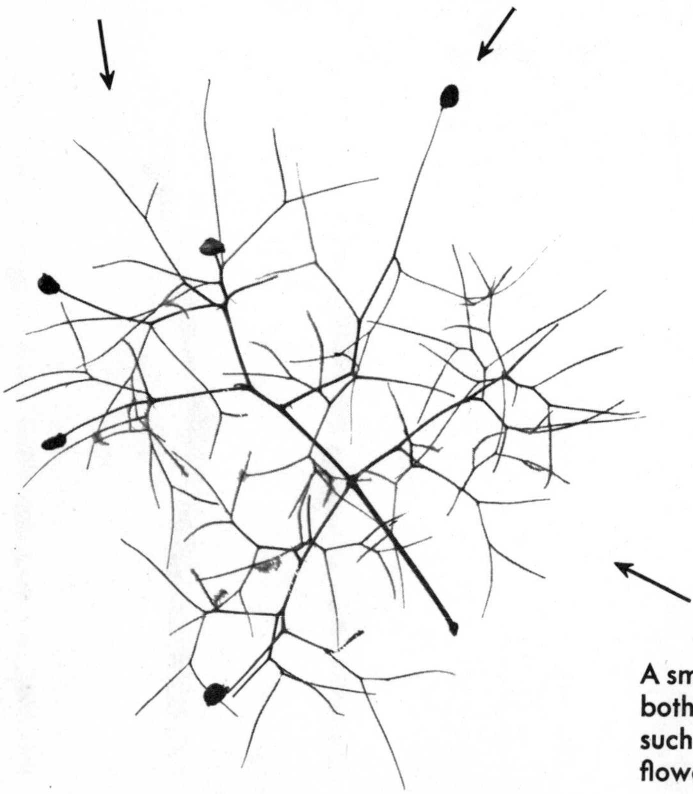

small stems
covered with
tiny hairs

actual fruit

**MP
72**

Note: These plants produce very few fertile
flowers. The individual flower stems of the
sterile flowers are covered with fine pink
or brownish hairs. These stems continue to
grow after flowering and become the
spectacular plumes that give these plants th
name of Smoke-trees. Notice that the
flower of American Smoke-tree (upper left)
shows the beginnings of the hairy stems, an
the picture of the fruiting plume (lower left)
shows these stems much elongated, with
only a few of them bearing actual fruit.

A small section of a large fruiting plume, typical of
both species. This developed from a flowering sectio
such as shown above at X. Bear in mind that both the
flowering and fruiting sections are actual size.

Smoke-tree

American
Smoke-tree

arks typical of both
pecies when small

American Smoke-tree

Smoke-tree

MP
73

SUMAC—*Rhus*   Dioecious or Polygamous   Alternate

Dwarf Sumac (Shining Sumac)	*R. copallina*	July-Aug.	Me., Ont. & Mich., s. to Fla. & Tex.	20 (occ. 30)
Fragrant Sumac	*R. aromatica*	March-April	Vt., s.w. Que. & Mich., s. to Fla. & Tex.	4 (occ. 6)
Poison Ivy	*R. radicans*	June-July	N.S. to Minn., s. to Fla. & Ariz.; vine, high-climbing support is available, or forming dense ground cov in open; occ. shrub-like on tops of stumps or wo	
Poison Sumac	*R. vernix*	June-July	Me. & Que. to Ont. & Minn., s. to Fla. & Tex.	20
Smooth Sumac	*R. glabra*	July-Aug.	Me. to B.C., s. to Fla. & Ariz.	10 (occ. 15)
Staghorn Sumac (Velvet Sumac)	*R. typhina*	June-July	Me., Que., Ont. & Minn., s. to Ga., Tenn., Ill. & Ia.	20 (occ. 30)

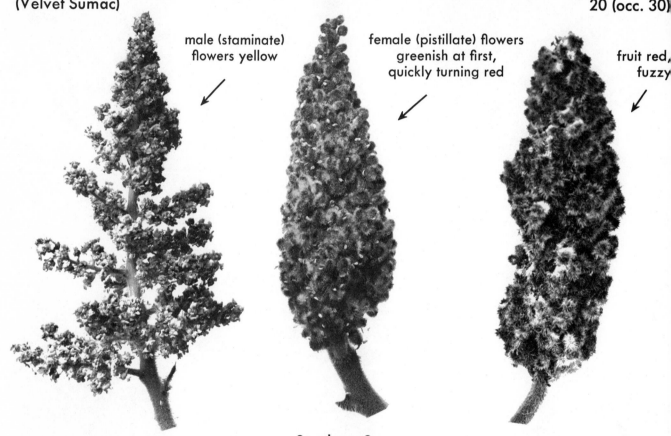

male (staminate) flowers yellow

female (pistillate) flowers greenish at first, quickly turning red

fruit red, fuzzy

Staghorn Sumac

flowers yellowish

fruit red, fuzzy

female (pistillate) flowers greenish (see Vine section, MP 183, for male flowers)

fruit dull; wh

Fragrant Sumac

Poison Ivy

male (staminate) flowers
yellow

female (pistillate) flowers
greenish, soon turning
red after pollinization

wers
eenish
ellow

Smooth Sumac
(Dwarf Sumac similar)

fruit red

Poison Sumac

ruit
dull white

son Sumac

Note: The flowers (and fruit)
of most Sumacs are terminal,
above the leaves, but those of
Poison Sumac and Poison Ivy
are along the stems, below
many of the leaves.

Smooth Sumac
(Dwarf Sumac similar)

MP
75

The three leaves at the top of this page are ½ actual size; the two leaves at the bottom are actual size.

Note wings along main leaf-stalk (rachis)
of Dwarf S.; Poison S. does not have these.

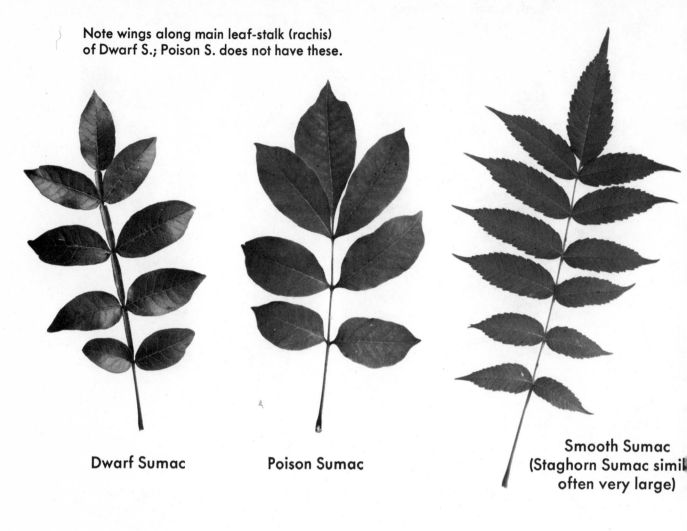

Dwarf Sumac

Poison Sumac

Smooth Sumac
(Staghorn Sumac simil
often very large)

Note: The end leaflet of the Poison Ivy
leaf is on a definite continuation of the
leaf-stalk beyond the two side leaflets.
The end leaflet of Fragrant S. begins
immediately above the two side leaflets.

aromatic
when crushed

Poison Ivy

Fragrant Sumac

fuzzy
twig

twigs pungently aromatic when crushed;
the only Sumac with winter catkins

ghorn   Smooth   Dwarf   Poison   Poison   Fragrant
umac   Sumac   Sumac   Sumac   Ivy   Sumac

barks typical of
Dwarf S.,
Fragrant S.,
Smooth S.,
Staghorn S. (only new
growth is fuzzy—see
twig, far left)

Note: Poison Ivy often climbs on trees and other objects by
means of rootlets, but
these are not always
present when the plant
is not climbing.

lenticels (dots on bark) generally
smaller than on other Sumacs

Poison Ivy

Poison Sumac

HOLLY—*Ilex* Dioecious Alternate

Black Alder *Ilex verticillata* (Winterberry)	June-July	Nfd. to Minn., s. to Ga. & Mo.	10-12 ft.
Inkberry *I. glabra*	June	mostly coastal, N.S. to Fla. & La.; evergreen 7-8 ft.
Large-leaved Holly *I. montana* (Mountain Winterberry; Mountain H.)	May-June	w. Mass. & N.Y., s. mostly on higher eleva- tions to S.C. & Ala. 25 ft.

female
(pistillate)
flowers

male
(staminate)
flowers

spurs

Black Alder

Large-leaved Holly

Inkberry

Note: Flowers and fruit are often on short spurs. These spurs, when present, distinguish this Holly at any time of year.

red
berries

red
berries

note spur

black
berries

MP
78

Black Alder

Large-leaved Holly

Inkberry

deciduous

evergreen

Black Alder

Large-leaved Holly

Inkberry

Black Alder

Large-leaved Holly

Inkberry

MP
79

EUONYMUS — *Euonymus* usually Perfect Opposite

Burning-bush *E. atropurpurea*   May-June N. Y. & Ont. to Minn., s. to Fla. & Te
(Wahoo)   20-25

European Spindle-tree *E. europaea*   May   Europe & w. Asia; escaped e. states 20

Running Strawberry-bush *E. obovata*   May   w. N. Y., s. Ont. & Mich., s. to Mc
(Running Euonymus)   Tenn., Mo.; trailing, rooting alor
                                              stems; tips ascending to 1 ft.

Strawberry-bush *E. americana*   May-June N. Y. to Mich. & Ill., s. to Fla. & Te
(Brook Euonymus; Hearts-a-burstin; Bursting Heart)   7-8

Winged Euonymus *E. alata*   May-June n.e. Asia & China; much planted
                                              escaped   10

Wintercreeper *E. fortunei*   June-July China; evergreen
   The two varieties below are more commonly encountered than the parent species:
      Common Wintercreeper *E. f. radicans* evergreen; trailing or climbing vine
      Bigleaf Wintercreeper *E. f. vegeta*   evergreen; high-climbing vine, sometimes somewh
                                             shrubby; heavier than Common Wintercreeper

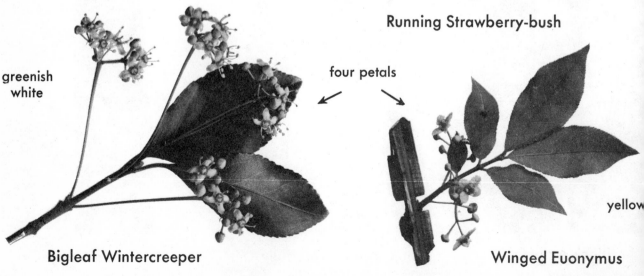

five petals

yellow
green

Strawberry-bush

purpl
gre

Running Strawberry-bush

greenish
white

four petals

Bigleaf Wintercreeper

yellow

Winged Euonymus

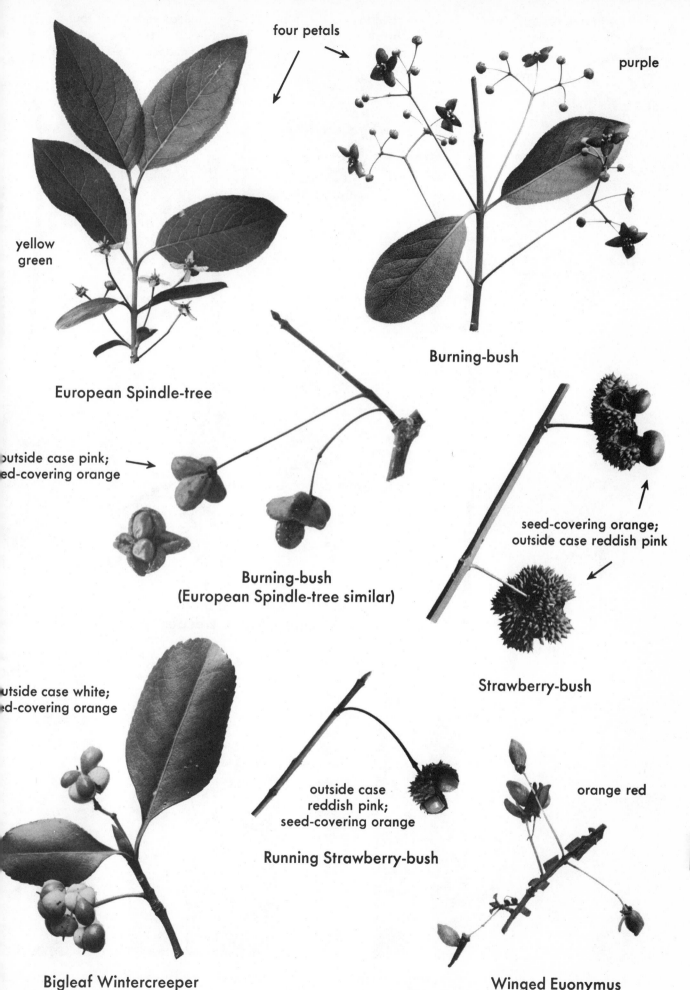

four petals

purple

yellow
green

Burning-bush

European Spindle-tree

outside case pink;
seed-covering orange

seed-covering orange;
outside case reddish pink

Burning-bush
(European Spindle-tree similar)

Strawberry-bush

outside case white;
seed-covering orange

outside case
reddish pink;
seed-covering orange

orange red

MP
81

Running Strawberry-bush

Bigleaf Wintercreeper

Winged Euonymus

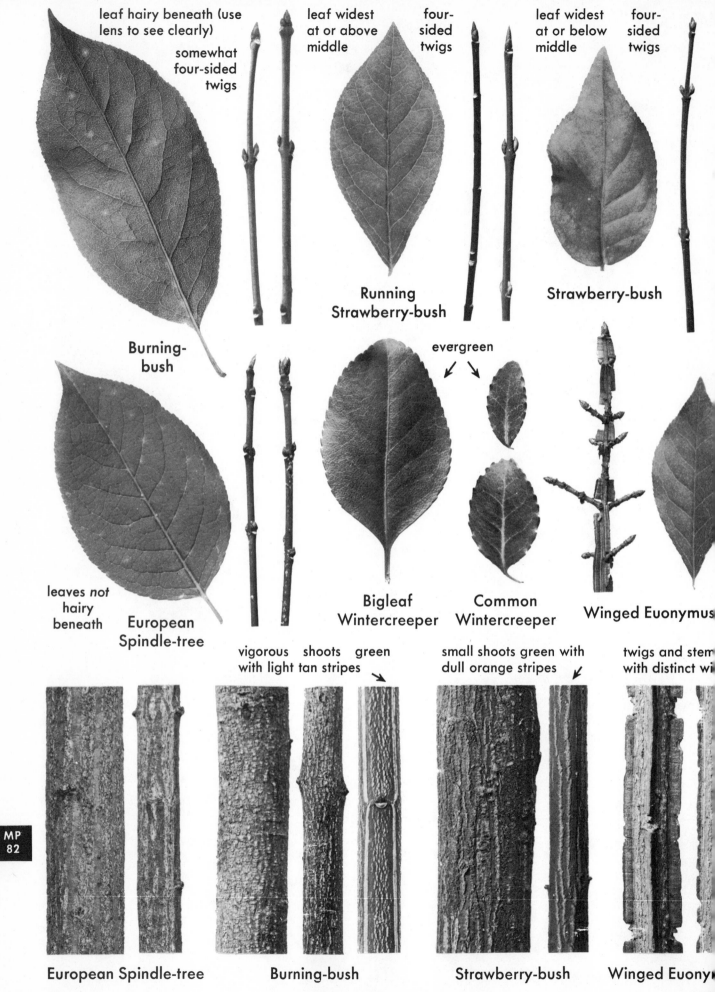

leaf hairy beneath (use lens to see clearly)

somewhat four-sided twigs

leaf widest at or above middle

four-sided twigs

leaf widest at or below middle

four-sided twigs

Burning-bush

Running Strawberry-bush

Strawberry-bush

evergreen

leaves *not* hairy beneath

European Spindle-tree

Bigleaf Wintercreeper

Common Wintercreeper

Winged Euonymus

vigorous shoots with light tan stripes

green

small shoots green with dull orange stripes

twigs and stem with distinct wi

European Spindle-tree

Burning-bush

Strawberry-bush

Winged Euony

MOUNTAIN HOLLY—*Nemopanthus mucronatus*

(not a true Holly; see MP 78)   Dioecious or Polygamous   May-June   Alternate
Nfd. to Wis. & Minn., s. to upland Va. & n. Ill.   10 ft.

female
(pistillate)
flowers

red berries

short, abrupt
tip (mucro)

male
(staminate)
flowers

purple
leaf-stalks

Note spurs on twigs
and small branches.

MP
83

BUCKTHORN—*Rhamnus* See individual species for kinds of flowers, and whether Opposite or Alternate

Alder-leaved Buckthorn	mostly Dioecious	May-June	Nfd. to B.C., s. to n. & w. N.E.,
Rhamnus alnifolia	Alternate		N.J., W. Va., n. Ia. & west 3 f
Common Buckthorn	mostly Dioecious	May-June	Europe & Asia; naturalized her
R. cathartica	mostly Opposite		sometimes tree-like 25 f
European Buckthorn	Perfect	May-July	Europe, w. Asia & n. Africa; na
R. frangula	Alternate		uralized here 20 f
(Alder Buckthorn)			
Lance-leaved Buckthorn	Perfect	May	locally distributed Pa. to Neb.,
R. lanceolata	Alternate		to Ala. & Tex. 6 f

Buckthorn flowers look alike superficially but note:

Alder-leaved B.,
    five stamens, five sepals
    and usually no petals

European B.,
    five stamens, five petals
    and five sepals

Common B. and
Lance-leaved B.,
    four stamens, four petal
    and four sepals

typical Buckthorn flowers

berries red to black

typical Buckthorn fruit

MP
84

250

Alder-leaved Buckthorn

European Buckthorn

Lance-leaved Buckthorn

Common Buckthorn

MP
85

Alder-leaved Buckthorn

Lance-leaved Buckthorn

side twigs often sharp-pointed

mostly opposite,
but occasionally
partly alternate

naked buds
(buds without scal

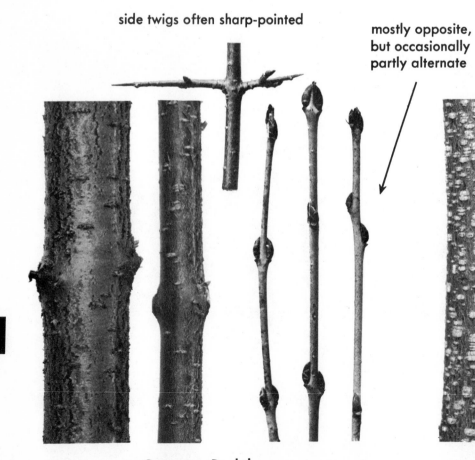

Common Buckthorn

European Buckthorn

AMERICAN BLADDERNUT—*Staphylea trifolia* Perfect   May   Opposite
N.H., Que. & Ont. to Minn., s. to
Ga. & Okla.   15 ft.

Seeds, when ripe,
rattle in papery pods.

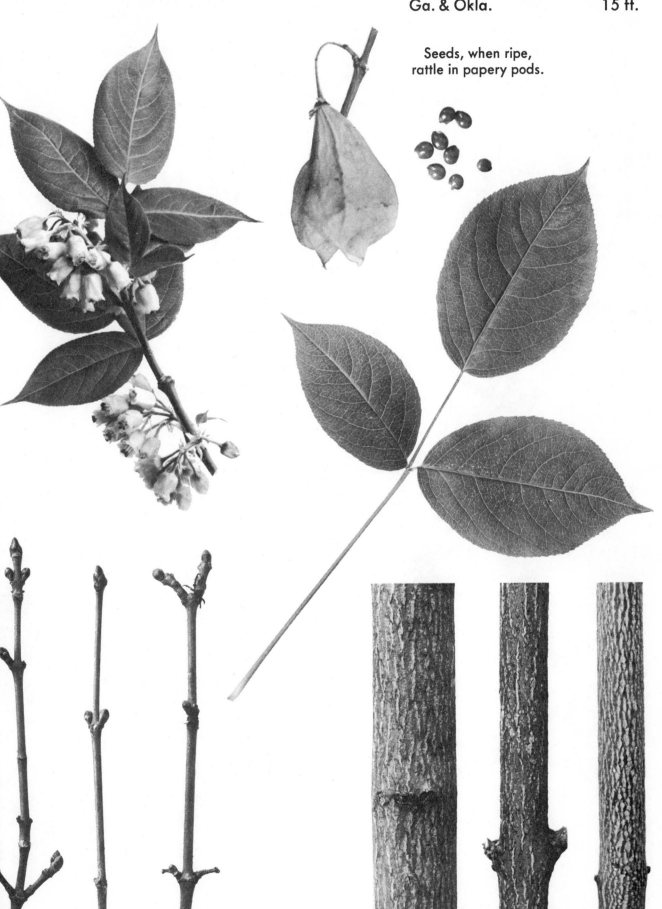

ST. JOHN'S-WORT—*Hypericum*   Perfect; five petals, five sepals   Opposite

Dense-flowered St. John's-wort	*Hypericum densiflorum*	July-Sept.	L.I. to W. Va., Mo., s. to Fla. Tex. 6 f
Shrubby St. John's-wort	*H. prolificum*	July-Sept.	N.Y., Ont. & Minn s. to Ga. & Ark. 5 f

yellow flowers

yellow flowers

Dense-flowered St. John's-wort

Shrubby St. John's-wort

T. PETER'S-WORT—*Ascyrum*   Perfect; four petals, four sepals in unequal pairs Opposite

t. Andrew's Cross   *A. hypericoides multicaule*   July-Sept.
Nantucket & L.I., Ky., s. Ind., s. Ill., Mo. & s.e. Kans. (for var. given here).
The standard species is a strictly southern shrub, more upright than this var.,
which has creeping stems with small branches ascending to only 6-8 in.

t. Peter's-wort   *A. stans*   July-Sept. L.I., Pa. & Ky., s. to Fla. & Tex.; upright shrub 2½-3 ft.

Note: The unequal sepals, the larger pair clasping as shown below, are distinctive of this genus.

Note markedly two-sided twigs (alternating flat sections along small stems) typical of both species.

yellow flowers

sepals removed to show seed pods

The flowers of St. Andrew's Cross are smaller, the petals narrower, than those of St. Peter's-wort, but are similar otherwise.

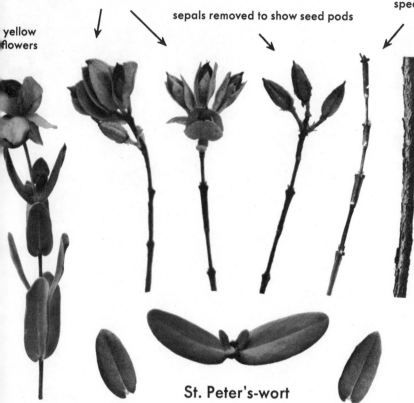

St. Peter's-wort

St. Andrew's Cross

ST. JOHN'S-WORT (con't.) typical bark and twigs

MP
89

ROSE OF SHARON—*Hibiscus syriacus*   Perfect   July-Sept.   Alternate
(Shrub Alth(a)ea)   China, India; much planted and escaped here
                                                 10 (occ. 15-20) ft.

Note: The flower of the standard type is single.
There are a number of garden forms with
double flowers.

The color varies
from white to
pink, red, blue
or purple.

LEATHERWOOD—*Dirca palustris*

(Moosewood, Wicopy, Leatherbush)

Perfect March-April Alternate
N.B., Ont. & Minn., s. to Fla. & La. 6 ft.

yellow flowers

green or red fruit

bark fibrous, pliable and strong; wood soft and weak

DAPHNE—*Daphne mezereum*

(Mezereum)

Perfect March-April Alternate
Europe; occ. escaped elsewhere 3 ft.

fragrant, pinkish-purple (occasionally white) flowers, long before other growth starts

red (or yellow) berries

MP 91

SARSAPARILLA or SPIKENARD — *Aralia*   Perfect   Alternate

Bristly Sarsaparilla   *Aralia hispida*   June-July   Nfd. to Man., s. to N.C., Ill. & Minn.; sub-
  (Dwarf Elder; Pigeonberry; Hyeble)   shrub woody only at base   3 f

Hercules' Club   *A. spinosa*
  (Devil's Walking-stick;   Aug.   s. N.Y. to s. Ill., s. to Fla. & Tex.; escape
  Toothache-tree; Angelica-tree)   elsewhere   35 f

flowers

½ actual
size

black
fruit

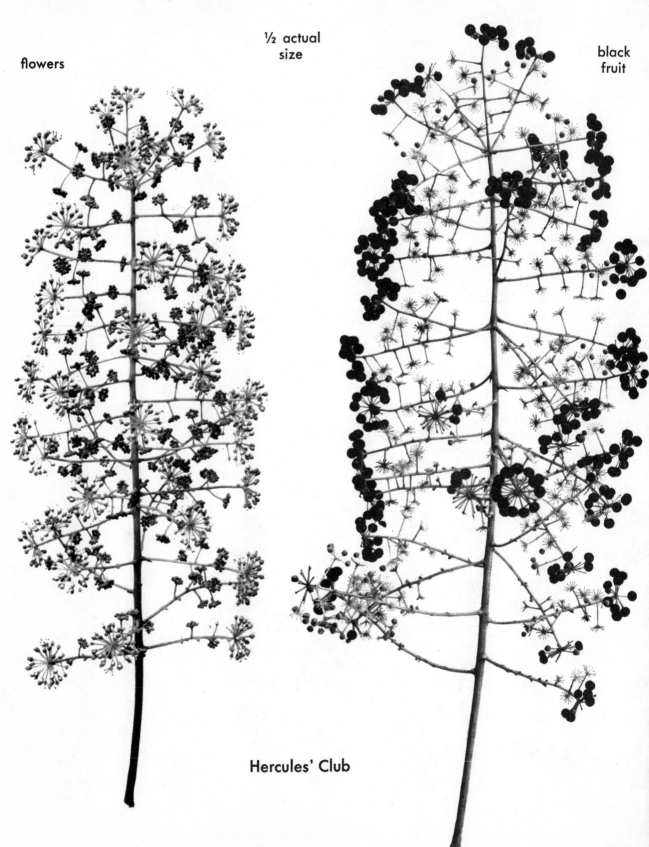

Hercules' Club

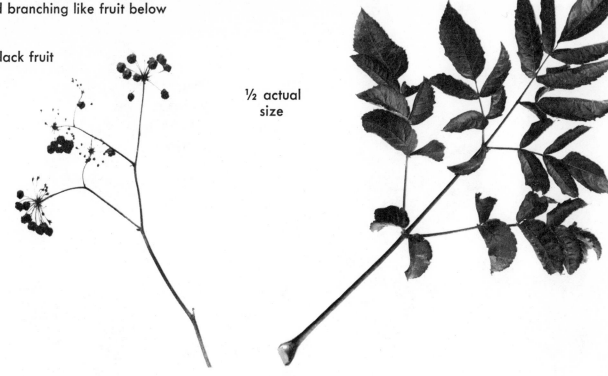

ristly Sarsaparilla flowers similar to
Hercules' Club, but smaller, less dense
and branching like fruit below

black fruit

½ actual
size

Bristly Sarsaparilla (see next page for
leaf of Hercules' Club)

usually, but not
always, thorny

actual
size

fruit remnants
sometimes found
in winter

MP
93

Hercules' Club

Bristly Sarsaparilla

This is all one compound leaf;
these leaves are often 3 feet long.

½ actual
size

Hercules' Club

UFFALO-BERRY— *Shepherdia canadensis* Dioecious April-May Opposite
(Canada Buffalo-Berry; Soapberry)   Nfd. to Alaska, s. to Me., w. Vt., n. & w. N.Y.,
n.w. Pa., n. Ohio, n.w. Ind., n.e. Ill., w. Ia. & west
6 (occ. 9) ft.

Note: Another Buffalo-berry—*S. argentea*—a western species, has silvery leaves, especially so beneath, and is often thorny, the twigs ending in sharp points (see Thorn Key, p. 1).

small yellowish
flowers

fruit yellow
to red

leaves slightly
silvery both sides

buds and twigs
somewhat silvery

typical Buffalo-berry barks

MP
95

OLEASTER—*Elaeagnus*   Perfect or Polygamous   Alternate

Autumn Elaeagnus	*E. umbellata*	May-June	e. Asia; escaped elsewhere	12 ft.
Russian Olive (Oleaster)	*E. angustifolia*	June	s. Europe, w. & c. Asia; escaped elsewhere	20-25 ft.
Silver-berry	*E. commutata*	May-June	Canada, s. to s. Minn., S. Dak. & Utah	12 ft.

typical Oleaster flowers

fruit red, dotted with silver or brown scales; juicy

twigs silvery or golden brown

leaves silvery beneath

Autumn Elaeagnus

fruit silvery, mealy

twigs golden brown

leaves silvery both sides, brown-dotted beneath

Silver-berry

fruit yellow or silvery, mealy

twigs silvery, fuzzy

leaves dull green above silvery beneath

Russian Olive

Autumn Elaeagnus and Russian Olive are sometimes thorny (see also Thorn Key, p. 2); Silver-berry has no thorns.

typical Oleaster barks

SWAMP LOOSE-STRIFE—*Decodon verticillatus*
(Water Willow)

Perfect   Aug.-Sept.   Opposite
Aquatic subshrub, woody only at base. The several-sided arching stems die back in winter except for tips, which root upon touching water or mud. In bogs and ponds, Me., s.w. Que., Ont. & Minn., s. to Fla. & La. Arching stems to 9 ft.

rose-purple flowers

leaves mostly in
whorls of three
around the stem

DOGWOOD—*Cornus*   Perfect   Opposite (except Alternate-leaved Dogwood)

Alternate-leaved Dogwood	*C. alternifolia*	May-June	Nfd. to Minn., s. to Ga., Ala. & Mo.	25
Bunchberry *C. canadensis*		May-July	Arctic, s. to Md., W. Va., Ill. & w. to Cali	
(Dwarf D.; Dwarf Cornel)			herbaceous	6-10
(see also Ground Covers, MP 169A)				
Flowering Dogwood *C. florida*		May	s. Me., s. N.H., s. Vt., N.Y., s. Ont., Ohio, Ind.	
			Ill., s. to Fla., Tex. & Mexico; often a tree 30	
Gray-stemmed Dogwood *C. racemosa*		June	Me. to Ont. & Minn., s. to Del., W. Va., n. Ind	
(Panicled D.; Gray D.)			Wis. & Ia. (possibly farther south) 6 (occ. 10)	
Red-Osier Dogwood *C. stolonifera*		May-June	Nfd. & west, s. to Md., Tenn., Ill., Ia. & Neb	
(Red-stemmed D.)			s.w. to N. Mex. & Calif.	6-8
Round-leaved Dogwood *C. rugosa*		May-June	Que. to Man., s. to s. N.E., Pa., W. Va. &	
				10 (occ. 15)
Swamp Dogwood *C. amomum*		June	Me. to Ind., s. Ga. & Ala.; western form, som	
(Red Willow; Swamp Cornel; Silky D.)			times called *C. obliqua*, extends range	
			N. Dak. & Okla.	10

Alternate-leaved
Dogwood (and other
Dogwoods; see Note, type 1)

Flowering Dogwood

Note: The flowers of all Dogwoods have
four petals. There are three types:
1. Alternate-leaved Dogwood. A flat
   cluster of many small flowers. Red-
   Osier, Round-leaved and Swamp
   Dogwoods have similar flowers.
2. Gray-stemmed Dogwood. Similar to
   type 1, except that the flower clusters are not flat, but more or less
   conical. This shape is also evident
   in fruit (see MP 100).
3. Flowering Dogwood and Bunchberry. The actual flower clusters of these two are small and relatively inconspicuous. They
   are placed in the middle of four large white (or pink) bracts,
   as can be seen above.

Bunchberry

Gray-stemmed Dogwood

small stems often red,
especially Swamp Dogwood

**Alternate-leaved Dogwood
(Swamp Dogwood similar)**

Flowering Dogwood

arge and small stems
d (there is also a yellow-
stemmed variety)

large and small stems
pinkish or greenish yellow

stems gray
(or brownish)

d-Osier Dogwood

Round-leaved Dogwood

Gray-stemmed Dogwood

MP
99

265

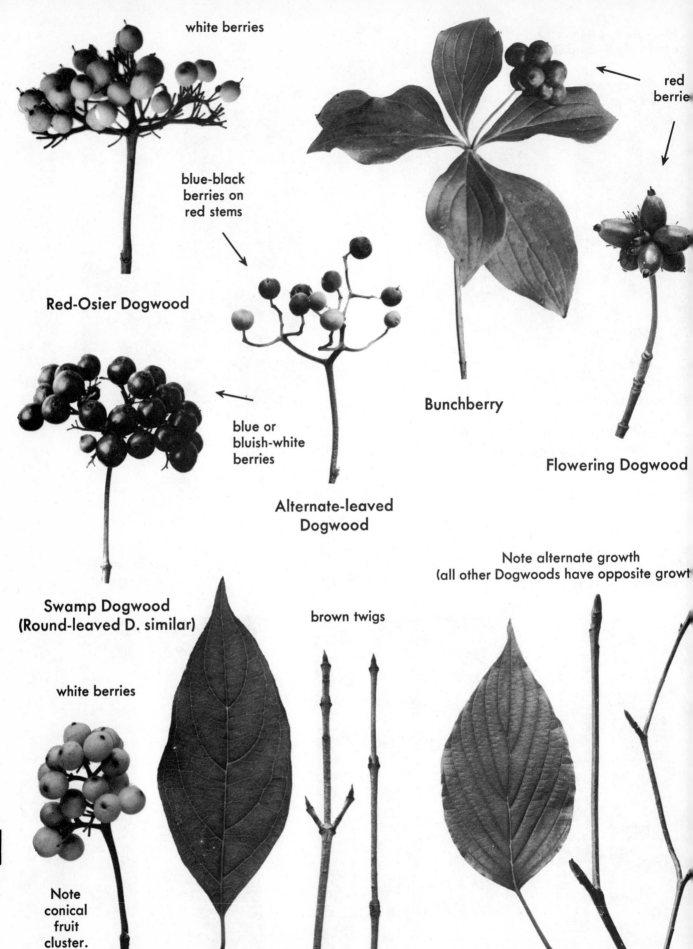

white berries

red berries

blue-black berries on red stems

Red-Osier Dogwood

blue or bluish-white berries

Bunchberry

Alternate-leaved Dogwood

Flowering Dogwood

Swamp Dogwood (Round-leaved D. similar)

Note alternate growth (all other Dogwoods have opposite growth

brown twigs

white berries

Note conical fruit cluster.

Gray-stemmed Dogwood

Alternate-leaved Dogwood

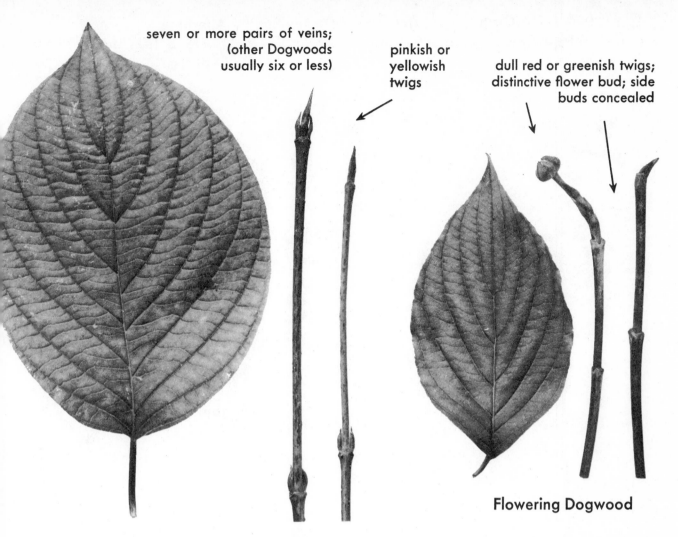

seven or more pairs of veins;
(other Dogwoods
usually six or less)

pinkish or
yellowish
twigs

dull red or greenish twigs;
distinctive flower bud; side
buds concealed

Flowering Dogwood

Round-leaved Dogwood

bright red (or maroon or green) twigs

brown
pith

leaf-stalk

Swamp Dogwood

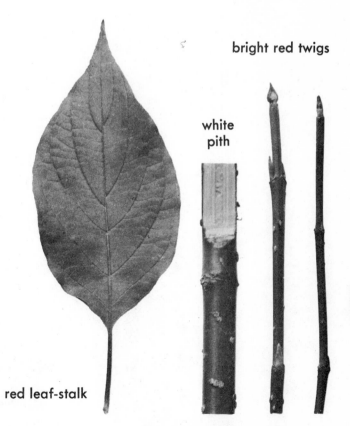

bright red twigs

white
pith

red leaf-stalk

Red-Osier Dogwood

MP
101

267

CLETHRA—*Clethra alnifolia*   Perfect   July-Sept.   Alternate
(Sweet Pepperbush; Summer Sweet; White-Alder)   Me. to Fla. & Tex., mostly near coas
9-10 ft

Clethra is one of the few plants to produce side growth from new shoots.

two types of bark, sometimes both on the same plant

MP
102

MENZIESIA — *Menziesia pilosa*   Perfect   May-June   Alternate
(Minniebush; Allegheny Menziesia)   Pa. to Ga. & Ala.   6 ft.

flowers greenish yellow, pped with d or pink

Note scattered whitish hairs on upper side and along edge of leaf (tip down and turn to see clearly).

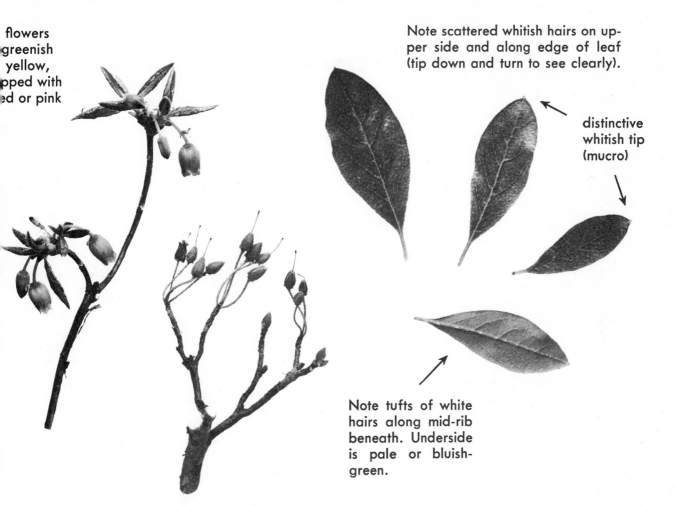

distinctive whitish tip (mucro)

Note tufts of white hairs along mid-rib beneath. Underside is pale or bluish-green.

bark rust colored; sometimes flaky, at other times smooth

twigs hairy; clusters of buds often at ends of twigs

AZALEA Subgenus I
RHODODENDRON Subgenus II — *Rhododendron* Perfect   Alternate, but note that all produce leaves in characteristic clusters at the ends of the twigs.

Note: Azaleas and Rhododendrons, although both in the genus *Rhododendron*, are usually placed in separate subgenera. Azaleas will be treated here first, as subgenus I, and Rhododendrons separately as subgenus II, on MP 108.

Subgenus I. AZALEA   Deciduous

Section 1.   Rhodora: ten stamens about as long as the petals, which are divided to the base of the flower (see opposite page)

Rhodora	*R. canadense*	April-May	Nfd. & Lab., s. to N.E., N.Y. & n.e. Pa. 3 ft.

Section 2.   Pentanthera: five stamens much longer than the petals, which are not divided all the way to the base of the flower

Flame Azalea	*R. calendulaceum*	May-June	s.w. Pa. to s.e. Ohio, s. to Ga. & Ala. 10 (occ. 15) ft.
Pink Azalea (Pinxterbloom; Election Pink; Wild Honeysuckle; Rose A.)	*R. nudiflorum*	April-May	s. Vt. to Ohio, s to S.C. & e. Tenn. 10 ft.
Rose Azalea (Downy Pinxterbloom; Election Pink; Honeysuckle A.; Mountain A.)	*R. roseum*	May-June	N.H. & s. Que. to Va. & Mo. 10 ft.
Swamp Azalea (Clammy A.; Swamp Honeysuckle)	*R. viscosum*	June-July	mostly in swamps, Me. to S.C., infrequently w. to Ohio & Tenn. 10 (occ. 15) ft.

There are a number of native Azaleas which in parts of their range appear to be distinct species but in other parts seem to merge into hybrid forms. As this is more confusing than important, except to the specialist, a few distinct types only are given here, with the suggestion that in general, the color, odor and form of the flowers, together with their blooming dates, are the most useful distinguishing characteristics, and that until considerable experience has been gained, identification of species should begin at flowering time. For example, Rose Azalea is almost identical in appearance with Pink Azalea. Typically, Rose Azalea blooms ten days to two weeks later than Pink Azalea, and has deeper pink, more fragrant flowers. The outside of the flower below the petals (called the tube) is sticky—a result of minute hairs ending in sticky glands (clearly visible under a lens). Pink Azalea has hairs on the tube but without the sticky glands. This constitutes an important difference, but also indicates how small the differences really are, and without the flowers it is often almost impossible to tell the two apart. In spite of these difficulties, one can always tell an Azalea from other plants at any time of year, and success with the species given here is fairly certain if attempted first at flowering time.

Subgenus II.   RHODODENDRON   Evergreen (see MP 108)

ubgenus I. AZALEA

Section 1. RHODORA: ten stamens about as long as the petals, which are divided to the base of the flower

pale to deep
rose purple

Rhodora
(April-May)

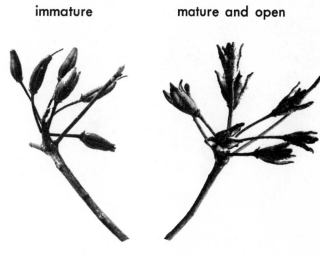

immature   mature and open

fruit typical of all Azaleas

Section 2. PENTANTHERA: five stamens much longer than the petals, which are not divided all the way to the base of the flower

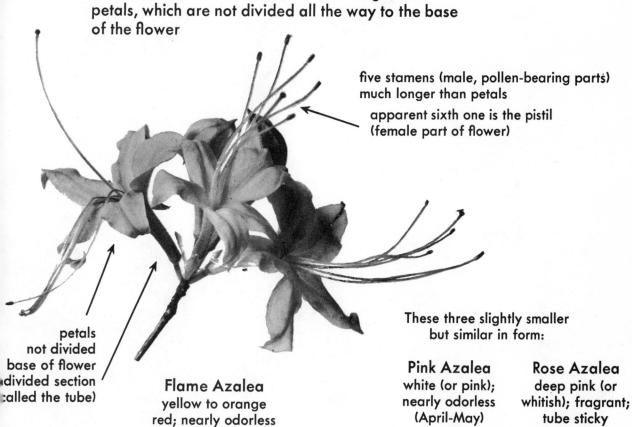

five stamens (male, pollen-bearing parts)
much longer than petals

apparent sixth one is the pistil
(female part of flower)

petals
not divided
base of flower
divided section
called the tube)

Flame Azalea
yellow to orange
red; nearly odorless
(May-June)

These three slightly smaller
but similar in form:

Pink Azalea
white (or pink);
nearly odorless
(April-May)

Rose Azalea
deep pink (or
whitish); fragrant;
tube sticky
(May-June)

Swamp Azalea
white;
fragrant
(June-July)

cluster of leaves at ends
of stems, typical of all
Azaleas (and Rhododendrons)

definite bluish cast above,
pale beneath

Rhodora

glossy, dark green

Swamp Azalea

These seven leaves are typical of Flame, Pink and Rose Azaleas; they are extremely variable in shape, size and texture.

MP
106

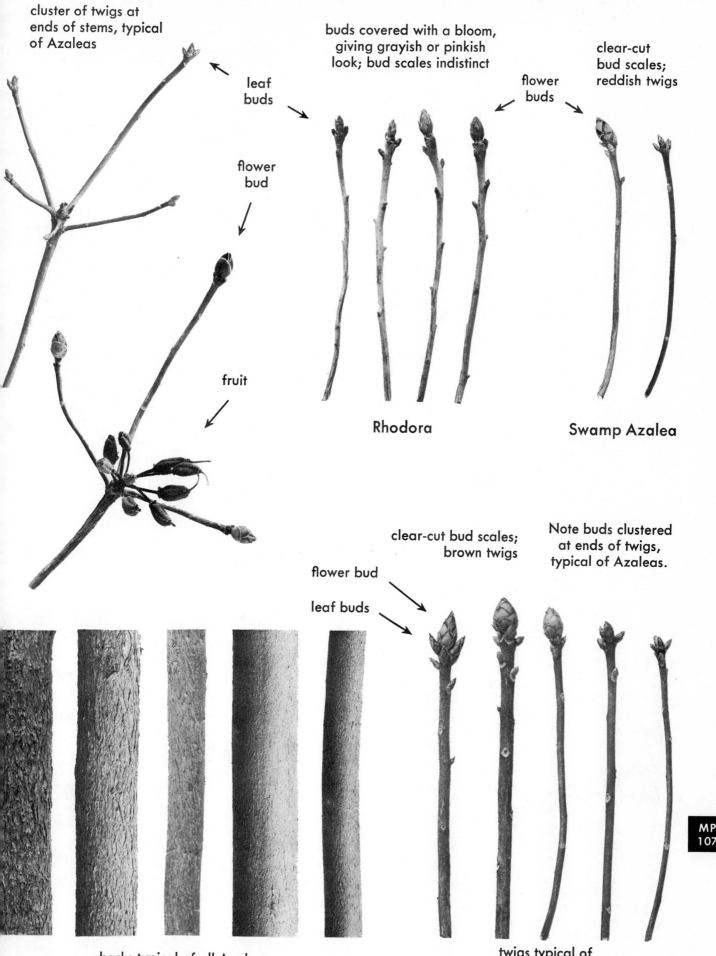

cluster of twigs at ends of stems, typical of Azaleas

leaf buds

flower bud

fruit

buds covered with a bloom, giving grayish or pinkish look; bud scales indistinct

flower buds

clear-cut bud scales; reddish twigs

flower buds

Rhodora

Swamp Azalea

clear-cut bud scales; brown twigs

flower bud

leaf buds

Note buds clustered at ends of twigs, typical of Azaleas.

barks typical of all Azaleas

twigs typical of Flame, Pink and Rose Azaleas

MP 107

Subgenus II. RHODODENDRON

Perfect Evergreen Alternate, but with leaves in clusters at ends of twigs

Carolina Rhododendron	*R. carolinianum*	May-June	N.C.; planted & hardy north 6-7 ft
Catawba Rhododendron (Mountain R. or Rosebay; Purple Laurel or R.)	*R. catawbiense*	May-June	W. Va. & Va., s. to Ga. & Ala. in mts.; planted & hardy north 6-7 (occ. 15) ft
Lapland Rhododendron (Lapland Rosebay; Rock Rose)	*R. lapponicum*	June-July	Alpine summits of n. Europe, n Asia & N. Amer.; dwarf Prostrate
Rosebay Rhododendron (Great Laurel)	*R. maximum*	June-July	n. N.E. to Ont. & Ohio, s. to Ga & Ala.; large-growing occ. to 35 ft

typical
Rhododendron
fruit

Rosebay Rhododendron
white to pink

Similar:
Carolina R.
  pink, smaller
Catawba R.
  lilac to purple

purple

leaves pitted
above; many
small brown
spots (scales)
beneath

typical Rhododendron barks

Lapland Rhododendron
(see also Ground Covers, MP 167

flower buds with leafy
appendages (bracts)

leaves whitish beneath; rounded at
both ends, especially at base

leaves green or
ownish beneath;
arply tapered at
both ends

flower buds without bracts

small leaf buds
covered by bracts

Catawba Rhododendron

leaves small;
usually rusty
beneath

Note: The leaves of
Rhododendrons are
largely clustered at
the ends of the stems
(as are those of
Azaleas; see MP 106).

MP
109

Rosebay Rhododendron

Carolina Rhododendron

LAUREL—*Kalmia* Perfect Evergreen See individual species for opposite or alternate characteristics.

Mountain Laurel (Spoonwood; Ivy)	*Kalmia latifolia*	May-July	Que. & N.B. to Ind., s. to Fla. & La.; usually alternate 15-20 (occ. 35) ft.
Sheep Laurel (Lamb-kill; Wicky)	*K. angustifolia*	May-July	Nfd. & Hud. Bay to Mich., s. to Ga.; usually opposite (or in 3's) 4-5 ft.
Swamp Laurel (Pale Laurel)	*K. polifolia*	May-July	mostly in bogs, Lab., Hud. Bay & west, s. to N.E., Pa., Ohio & Wis.; usually opposite (or in 3's) 2-2½ ft.

white to pink

deep pink

Mountain Laurel

Sheep Laurel

pink

Swamp Laurel

Note: Mountain Laurel } flowers and fruit
      Swamp Laurel } terminal (at end of stem)

Sheep Laurel — flowers and fruit
      lateral (along stem
      and below newest
      growth)

MP
110

Mountain Laurel

Swamp Laurel

Sheep Laurel

Note distinctive flower
buds, typical of Laurel.

dark, glossy green above,
yellow green beneath

green above, whitish
beneath; edges of lea
curl under (revolute)

Mountain Laurel

Swamp Laurel

light green above,
pale or whitish beneath

typical Laurel bark

Sheep Laurel

MP
112

OG ROSEMARY—*Andromeda glaucophylla*
(Downy Bog Rosemary)

Perfect   May-July   Evergreen   Alternate
mostly in bogs, Nfd. & Lab. to Man., s. to N.E.,
N.J., W. Va., n.e. Ill. & Minn. 1 ½ (rarely 2 ½) ft.

owers white,
nged with pink

ote:
wamp Laurel
opposite
age is somewhat
milar but has
pposite leaves
ith no bluish cast.

Mature leaves have a bluish
cast above and are distinctly
white below. The edges curl
under (revolute).

AND-MYRTLE—*Leiophyllum buxifolium*

Perfect   May-June   Evergreen   mostly Alternate
Pine Barrens of N.J. & in var. s. to S.C., Tenn. & Ky.
1 ½ (rarely 2 ½) ft.

ANDROMEDA—*Pieris* Perfect Evergreen Alternate (rarely Opposite)

Fetter-bush	*Pieris floribunda*	April-May	Va. & W. Va., s. to Ga. & Tenn.	6 ft.
(Mountain Andromeda)				
Japanese Andromeda	*P. japonica*	April-May	Japan; much planted here	9-10 ft

flowers erect

Fetter-bush

flowers drooping

Japanese Andromeda

typical Andromeda ba

fruit drooping

fruit erect

Japanese
Andromeda

flower buds

Fetter-bush

Japanese
Andromeda

flower buds

tapered
at base

rounded
at base

Fetter-bush

LABRADOR TEA—*Ledum groenlandicum* Perfect May-June Evergreen Alternate
Greenland & Alaska, s. to N.E., n. Pa., Ohio, Mich., Wis., Minn. & west 3 ft

Note the unusual way the fruit
opens, hinging from the outer end.

leaves alternate but often
in tight clusters

Stems and undersides of leaves
are very woolly. This wool is at first
green, later rusty orange. The
leaves curl under at the edges
(revolute).

LEATHERLEAF—*Chamaedaphne calyculata* Perfect April-June Evergreen Alternate
(Dwarf Cassandra) n. Europe & n. Asia; N. Amer.: s. near coast to Ga., w. from Pa. to Ind., Wis.,
Minn. & B.C.
4 ft

leaves covered with
many small dots

| Male-berry | *Lyonia ligustrina* | May-June | Me. to Mich., s. to Fla. & Tenn.; in var. w. to Okla. & Tex.   12 ft. |
| tagger-bush | *L. mariana* | May-June | R.I. to Fla., w. to Tenn., Ark. & Tex.   6-8 ft. |

Male-berry   Stagger-bush

LEUCOTHOE or FETTER-BUSH—*Leucothoë*   Perfect   Alternate

Drooping Leucothoë *Leucothoë catesbaei* (Dog Hobble; Switch-ivy)	April-May	Va., Ga. & Tenn.; evergreen 7 ft.
Swamp Leucothoë *L. racemosa* (Sweet Bells; Fetter-bush)	May-June	Mass. & N.Y., s. near coast to Fla., w. to Tenn. & La.; deciduous 12 ft.

deciduou

evergreen

twig with all but
one leaf removed
to show
flower buds

flower
buds

flow
bud

Drooping Leucothoë

Swamp Leucothoë

MP
118

BLUEBERRY—*Vaccinium* Perfect Alternate
(Including Bilberry, Blueberry, Cranberry, Deerberry)

The Blueberries include a number of subgenera such as those shown above. For the sake of simplicity, the genus is divided here into three easily distinguishable groups:

I. BLUEBERRIES and DEERBERRY Deciduous; upright shrubs; berries blue, black, purple or green

Black Highbush Blueberry	*V. atrococcum*	May	Me. to Ont. & Mich., s. to Fla. & Ark. 10 (occ. 15) ft.
(Blue) Highbush Blueberry	*V. corymbosum*	May	Me. to Mich., s. to Fla. & La. 10 (occ. 15) ft.
Deerberry *V. stamineum* (Squaw Huckleberry)		May-June	Mass. to s.e. Mich., s. to Fla. & La. 3-5 ft.
Early Lowbush Blueberry) (Low Sweet Blueberry)	*V. vacillans*	May-June	Me. to Mich., s. to Ga. & Mo. 2 (occ. 3) ft.
Late Lowbush Blueberry *V. angustifolium* (Dwarf Sweet Blueberry)		May-June	Nfd. to Sask., s. to mts. of N.H. & N.Y., w. to Minn. & in var. farther south; 1-1½ ft., prostrate on higher mts.

II. MOUNTAIN CRANBERRY and CRANBERRIES (MP 124)   Evergreen; low or prostrate; berries red

Mountain Cranberry *V. vitis-idaea minus* (Rock C.; Cowberry; Lingenberry) (not a true Cranberry)		May-June	Lab. to Alaska, s. to Mass., Minn. & B.C.; creeping stems to 12 in., prostrate at higher elevations
Large Cranberry *V. macrocarpum*		June-Aug.	Nfd. to Sask., s. to N.C., Tenn., Ill., Wis. & Minn.; mostly in bogs & wet ground; creeping stems, slightly heavier than next, the tips more ascending
Small Cranberry *V. oxycoccus*		May-July	N. Amer.: s. to N.E., N.J., W.Va., Ill., Wis., Minn. & west; mostly in bogs & wet ground; stems creeping

III. BILBERRIES (MP 125) Deciduous; low or prostrate; berries blue or purple

Bog Bilberry *V. uliginosum*		May-June	n. Eu. & n. Asia; N. Amer.: s. to n. N.E., n. N.Y., Mich. & Minn., mostly in mts. (rarely to sea level); rarely to 1½ ft., usually nearly prostrate
Dwarf Bilberry *V. cespitosum*		May	N. Amer.: s. to n. N.E., n. N.Y., Mich., Minn., Colo. & n. Calif.; in alpine areas but also to sea level; rarely to 1 ft., usually in tufts only a few in. high

Like a few other genera included in this book, the Blueberries are an unstable group undergoing evolutionary changes, which makes species identification uncertain in many cases. The selection given above as representative of the main types was intentionally limited to emphasize the futility of any attempt at final definition, until such time as the group gives evidence of greater stability. The various dubious species and innumerable forms can safely be left to the dedicated hair-splitters.

I. BLUEBERRIES and DEERBERRY

Note: Both the Blue and Black Highbush Blueberries bloom at about the same time. The leaves of the Blue Highbush are then fairly well developed, whereas those of the Black Highbush develop after the flowers.

white flowers
with some pink
or red parts

undeveloped
leaves

(Blue) Highbush Blueberry

Black Highbush Blueberry

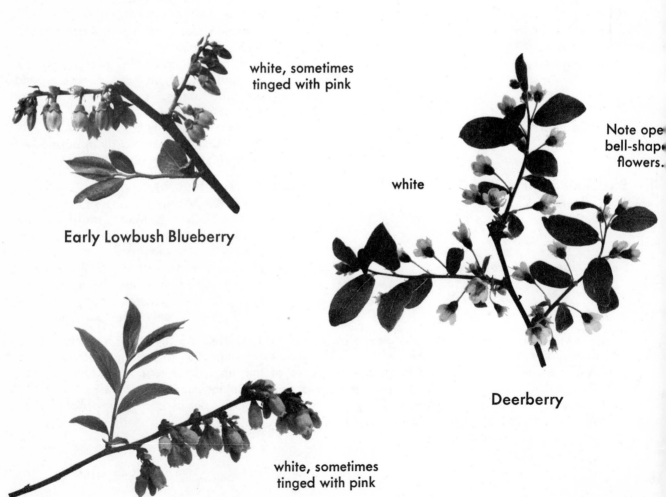

white, sometimes
tinged with pink

Early Lowbush Blueberry

white

Note ope
bell-shap
flowers.

Deerberry

white, sometimes
tinged with pink

Late Lowbush Blueberry

typically smooth or slightly hairy beneath; green or yellow green above

typically woolly beneath; glossy dark green above

blue with bloom

(Blue) Highbush Blueberry
(Black Highbush B. similar, but with glossy black berries)

lue) Highbush Blueberry

Black Highbush Blueberry

blue with bloom

Early Lowbush Blueberry
(Late Lowbush B. similar)

(Blue) Highbush Blueberry
(Black Highbush B. similar, except as noted above)

greenish (or slightly purplish) with bloom

Deerberry
(note characteristic small leaves on fruiting branches)

Early Lowbush Blueberry

leaves with distinctive bluish cast above; pale beneath

Late Lowbush Blueberry

Deerberry

MP 121

twigs reddish or green; smooth; not dotted

twigs reddish or green; larger ones, especially, covered with many minute dots

twiggy look, typical of all Blueberries

twigs numerous; mostly very fine

Early Lowbush Blueberry

Late Lowbush Blueberry

reddish or green

reddish, usually with some bloom

MP 122

Highbush Blueberry
(both Black and Blue)

Deerberry

Blueberries have extremely variable but distinctive barks. The Huckleberries, which are often confused with Blueberries, have uniform gray bark (see MP 127).

The older stems of Blueberries have rough, flaky bark; the young growth is smooth: maroon, red, green, yellow or brown, often all on the same plant.

Lowbush Blueberry (both Early and Late)

Highbush Blueberry (both Black and Blue)
(Deerberry similar)

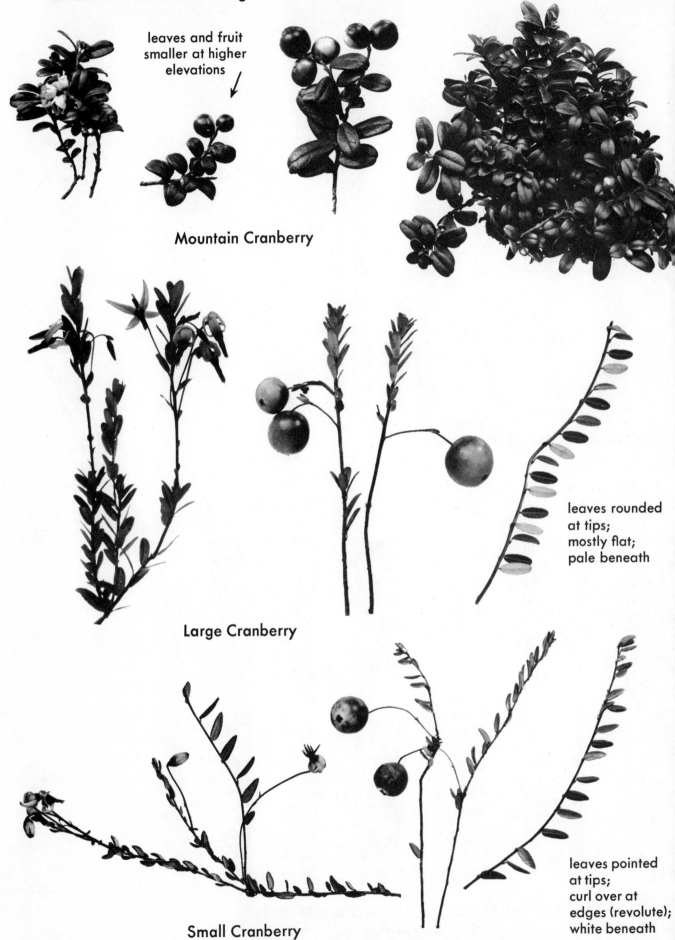

II. MOUNTAIN CRANBERRY and CRANBERRIES   Evergreen; all have red berries, sometimes white before maturing

leaves and fruit
smaller at higher
elevations

Mountain Cranberry

Large Cranberry

leaves rounded
at tips;
mostly flat;
pale beneath

Small Cranberry

leaves pointed
at tips;
curl over at
edges (revolute);
white beneath

III. BILBERRIES Deciduous; berries blue or purple with bloom

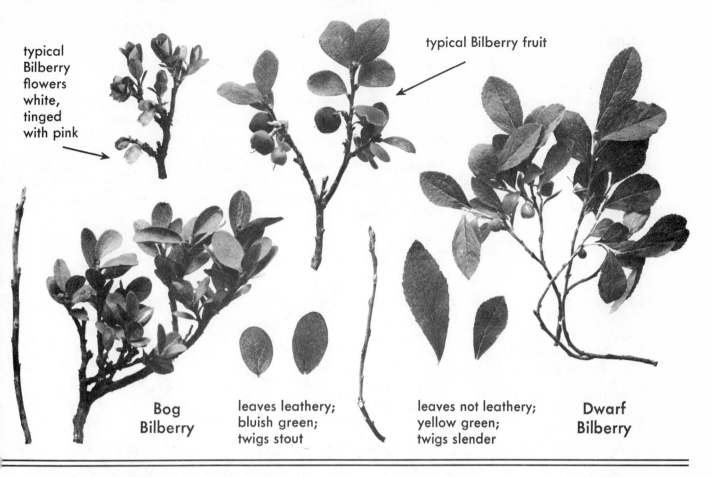

typical Bilberry flowers white, tinged with pink

typical Bilberry fruit

Bog Bilberry

leaves leathery; bluish green; twigs stout

leaves not leathery; yellow green; twigs slender

Dwarf Bilberry

COMMON MATRIMONY-VINE—*Lycium halimifolium*

Note: The Chinese Matrimony-vine—*L. chinense*—is similar but usually thornless and has wider leaves.

Perfect   May-Sept.   Alternate
s.e. Europe & Asia; escaped elsewhere; rambling shrub, sometimes supported on trellises or against buildings   10 ft.

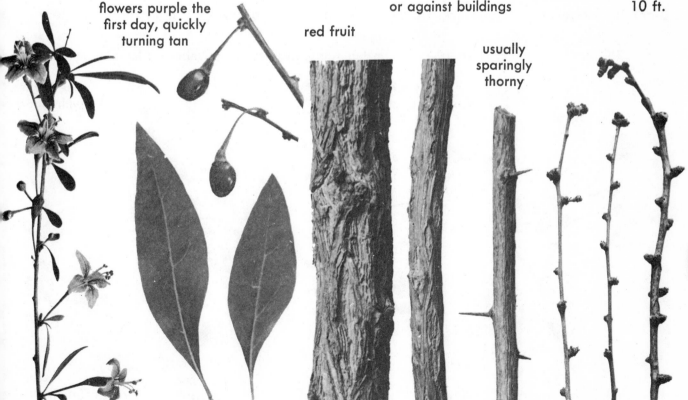

flowers purple the first day, quickly turning tan

red fruit

usually sparingly thorny

MP 125

291

HUCKLEBERRY — *Gaylussacia*   Perfect   Alternate

Black Huckleberry *G. baccata* (Highbush Huckleberry)	May-June	Nfd. to Man. & e. Minn., s. to Ga. & La.	3 ft.
Blue Huckleberry *G. frondosa* (Dangleberry, Tangleberry)	May-June	N.H. to La. near coast	6-7 ft.
Dwarf Huckleberry *G. dumosa* (Swamp Huckleberry)	May-June	Nfd. to La. near coast, infrequent in- land to W. Va. & Tenn.	1 ½ ft.
Evergreen Huckleberry *G. brachycera* (Box-Huckleberry)	April-May	Del. & Pa. to Tenn.	1 ½ ft.

Note: Both Blue and Dwarf Huckleberries have leaf-like bracts on the flower
stems; bracts of Dwarf Huckleberry persist into fruit.

flowers white
tinged with p[i]
or red

flowers mostly
reddish

flowers white,
tinged with
pink

true
leaf

bracts

Black Huckleberry   Blue Huckleberry   Dwarf Huckleberry   Evergreen Huckleberry

Note: The berries of all Huckleberries contain ten fairly large seeds. (Blueberries have many small seed[s]

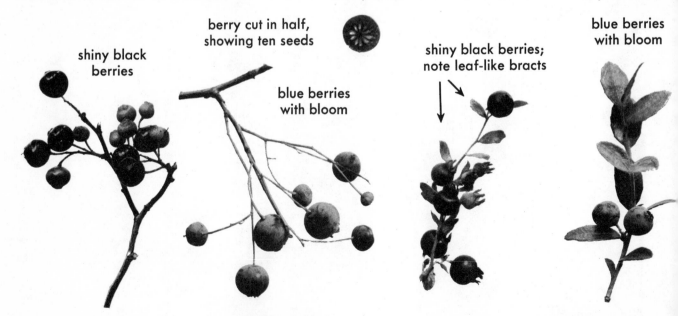

shiny black
berries

berry cut in half,
showing ten seeds

blue berries
with bloom

shiny black berries;
note leaf-like bracts

blue berries
with bloom

Black Huckleberry   Blue Huckleberry   Dwarf Huckleberry   Evergreen Huckleb[erry]

Note: The deciduous Huckleberries have yellow resin dots on the leaves (at least, always on underside). These are difficult to see without a magnifying glass, but are very useful in case of doubt.

green or yellow green above

somewhat bluish green above

dark green and glossy above, with distinct bristle tip (mucro)

pale beneath

Black Huckleberry

Blue Huckleberry

Dwarf Huckleberry

Evergreen Huckleberry

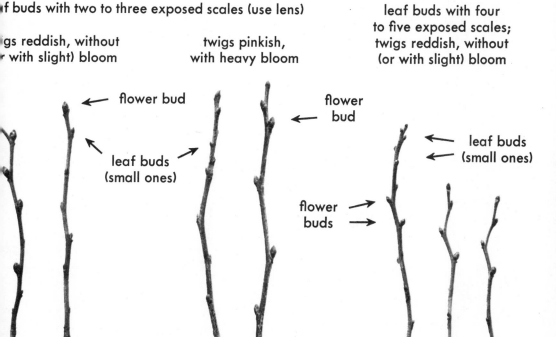

...f buds with two to three exposed scales (use lens)

...igs reddish, without ...r with slight) bloom

twigs pinkish, with heavy bloom

leaf buds with four to five exposed scales; twigs reddish, without (or with slight) bloom

flower bud

leaf buds (small ones)

flower bud

leaf buds (small ones)

flower buds

...ck Huckleberry

Blue Huckleberry

Dwarf Huckleberry

typical Huckleberry bark (fairly uniform throughout)

SILVERBELL—*Halesia carolina* Perfect April-May Alternate
(Snowdrop Tree; Opossum-wood; Shittimwood) W. Va. to Ill., s. to Fla. & Tex. 35 ft.

FRINGE-TREE—*Chionanthus virginicus*

Dioecious (or functionally so) May-June Opposite (occ. partly Alternate)
N.J., e. Pa. & W. Va., s. to Fla. & Tex.   30 ft.

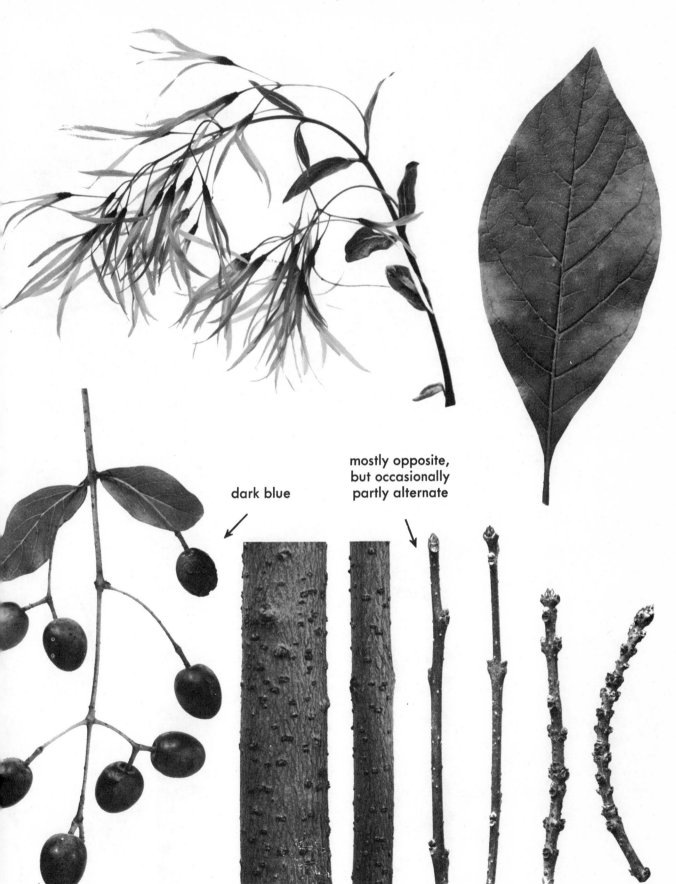

dark blue

mostly opposite,
but occasionally
partly alternate

MP
129

COMMON LILAC—*Syringa vulgaris*   Perfect   April-May   Opposite
(Old-fashioned Lilac)   There are many Lilac species and hybrids, all originating in s.e. Europe & Asia, but this is the only one frequently escaping in this country.   20 ft.

flowers purple before opening, pale lavender when open and very fragrant (also a white variety)

bark shaggy when larger

FORSYTHIA—*Forsythia* Perfect April-May Opposite
(Golden-bells) China 10 ft.

There are a number of species and hybrids. The following are the standard types:

Weeping Forsythia *F. suspensa*	pendulous branches; hollow stems except solid at nodes (joints); two kinds of leaves (see below)
Greenstem Forsythia *F. viridissima*	upright branches; stems with chambered pith; simple leaves only (bottom type)
Border Forsythia *F. intermedia*	a cross between the above two

all with yellow flowers

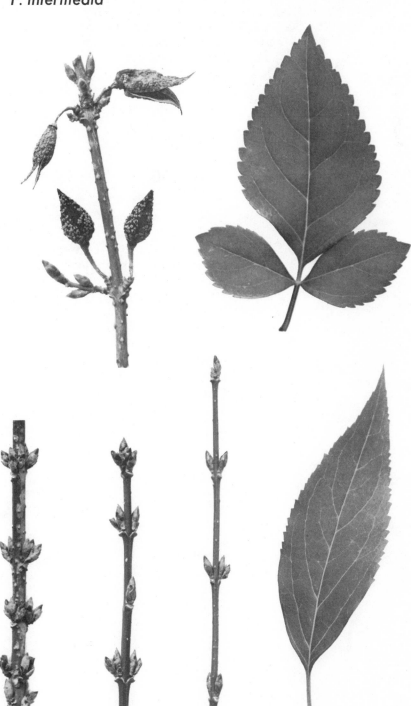

PRIVET—*Ligustrum*   Perfect   Opposite

California Privet	*L. ovalifolium*	July	Japan; escaped elsewhere		15 f
Common Privet	*L. vulgare*	June	Europe & N. Africa; escaped elsewhere		15 f
Ibota Privet	*L. obtusifolium**	June	Japan; escaped elsewhere		10 f

(Regal Privet—*L. o. regelianum*—is a dwarf variety, not over 3-4 ft.)
*The name Ibota is commonly used for this Privet although more correct-
ly applied to another species—*L. ibota.*

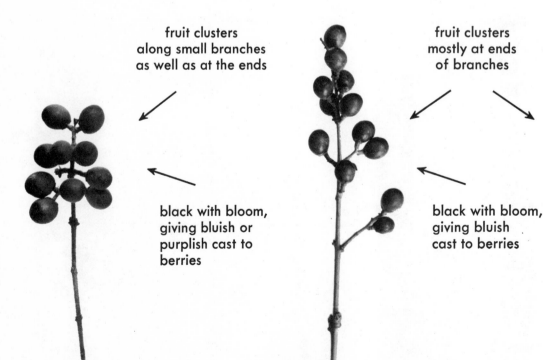

short petals
(less than half total
length of individual
flower); flower
clusters small,
produced all along
stems, not just at ends

long petals
(at least half total
length of individual
flower); flower
clusters large,
mostly only at
ends of branches

Note: Ibota and Common P. bloom at
about the same time. California P.
blooms two to three weeks later. It has
short petals like Ibota P., but the flower
clusters are large and at the ends of the
stems like those of Common P.

Ibota Privet

Common Priv

fruit clusters
along small branches
as well as at the ends

fruit clusters
mostly at ends
of branches

gloss
black

black with bloom,
giving bluish or
purplish cast to
berries

black with bloom,
giving bluish
cast to berries

Ibota Privet   **California Privet**   **Common Privet**

California Privet

Common Privet

Ibota Privet

California Privet

Common Privet

Ibota Privet

MP
133

ELDER or ELDERBERRY—*Sambucus*   Perfect   Opposite

Common Elder   *Sambucus canadensis*   June-July   N.S. to Man., s. to Fla. & Tex.   12 ft
   (American Elder)
Red-berried Elder   *S. pubens*   April-May   Nfd. to Alaska, s. to Ga., Tenn., Ind
   (Scarlet Elder)   la. & west   25 ft

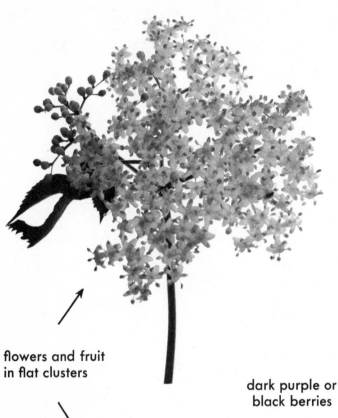

flowers and fruit
in flat clusters

dark purple or
black berries

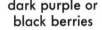

red berries

Common Elder   Red-berried Elder

typical
Elder leaf

Common Elder:
green leaf-stalk

Red-berried Elder:
purplish leaf-stalk

white pith

brown pith

Common Elder

Red-berried Elder

VIBURNUM— *Viburnum* Perfect Opposite

Arrow-wood *V. dentatum*	May-June	N.B. to Mich., s. to Ga. & Tenn.	15
Black-haw *V. prunifolium* (Stag-bush, Sweet-haw)	April-May	Conn. to Mich., s. to Fla. & Tex.	15
Downy Viburnum *V. rafinesquianum* (Downy Arrow-wood)	May-June	N.H. & Que. to Minn., s. to Ga. & Ark.	6
Highbush Cranberry *V. trilobum* (Pimbina)	May-July	N.B. to B.C., s. to N.E., Pa., W. Va., Ill., Ia. & west	12
Hobble-bush *V. alnifolium* (Witch-hobble; Tangle-legs; Moosewood)	April-June	N.B. to Mich., s. to Ga. & Tenn. stems often arching & rooting at tips	10
Maple-leaved Viburnum *V. acerifolium* (Maple-leaved Arrow-wood; Dockmackie)	May-June	N.B. to Mich. & Wis., s. to Ga. & Tenn.	6
Nanny-berry *V. lentago* (Sheep-berry; Sweet Viburnum; Wild Raisin)	May-June	Me. & Que. to Man., s. to Ga., Ill. & Colo.	30
Smooth Withe-rod *V. nudum* (Possum-haw; Swamp-haw)	June-July	Conn. to e. Tex. & Fla.	15
Southern Arrow-wood *V. pubescens*	June-July	Mass. to Pa., s. to Fla. & Tex.	10
Withe-rod *V. cassinoides*	June-July	Nfd. to Ont., Mich. & Wis., s. to. N.C. & Ala.	10

Note: There are two main types of
   flowers:

I. with fertile flowers only

II. with large marginal, sterile
   flowers, and small fertile ones
   forming the central flower
   cluster

The flowers of
all Viburnums have
five petals.

Type I
Arrow-wood
Black-haw
Downy Viburnum
Maple-leaved Viburnum
Nanny-berry
Smooth Withe-rod
Southern Arrow-wood
Withe-rod

Type II
Highbush Cranberry
Hobble-bush

MP
136

red (not always so large)

red, eventually
turning dark
purple or black

Hobble-bush

yellow to
pink, finally
dark blue
with bloom

Highbush Cranberry

dark blue

pink, turning dark blue,
but with a heavy bloom
giving berries light blue
appearance

Nanny-berry
(Black-haw similar)

Arrow-wood
(Southern A. similar)

glossy black,
sometimes with a
bloom giving berries
bluish look

considerable variation
in size and shape,
me berries being flattish,
others round

MP
137

Withe-rod
(Smooth W. similar)

sometimes more oval
or flattened, like
Maple-leaved Viburnum at
extreme left

Maple-leaved Viburnum
(Downy V. similar)

usually
very
large

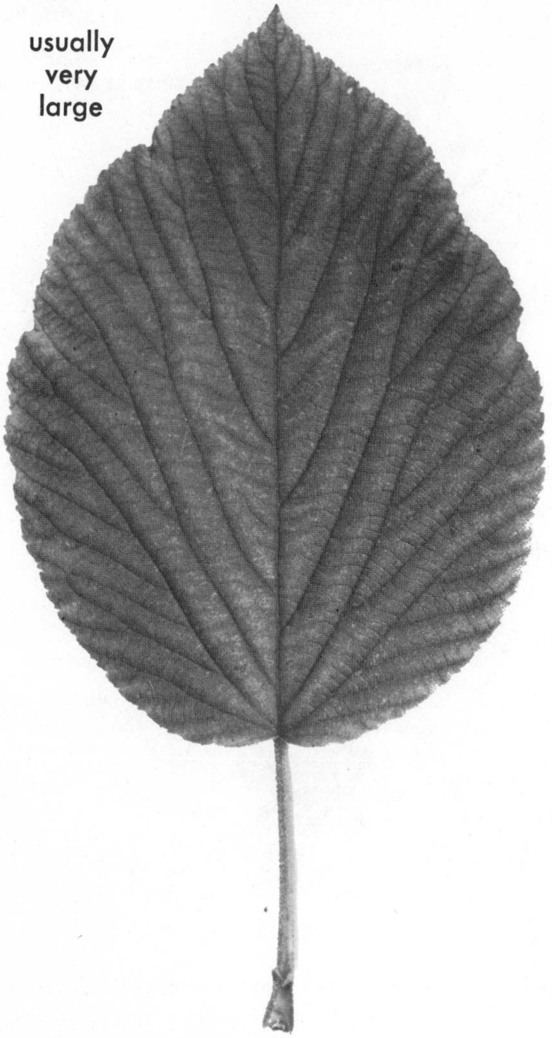

Hobble-bush

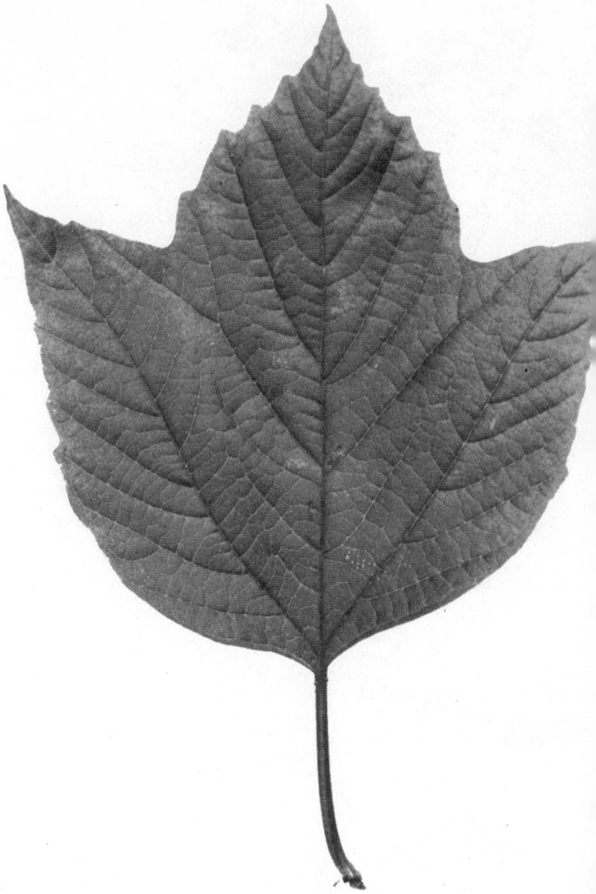

Highbush Cranberry

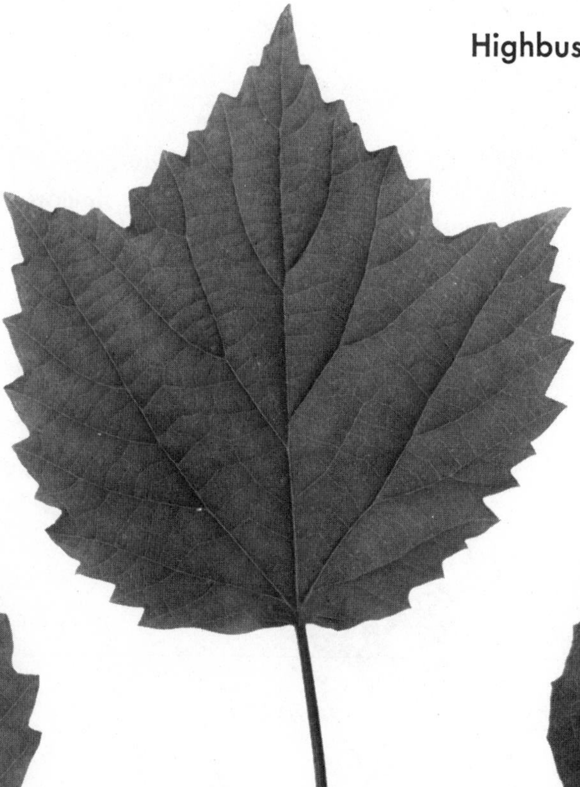

The center leaf is th
mature form, but th
two smaller ones a
often found on your
shoots or new growt

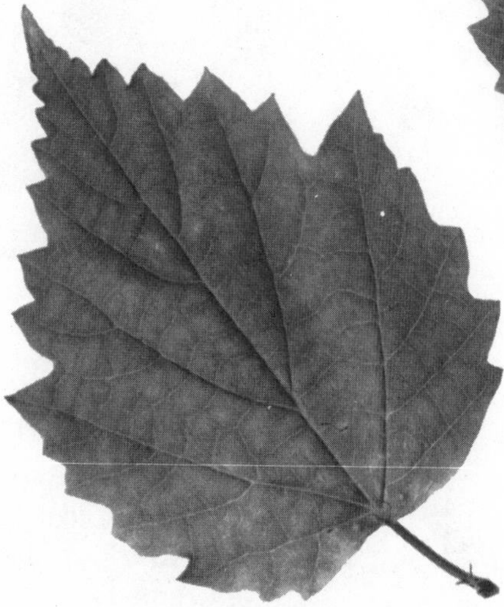

MP
138

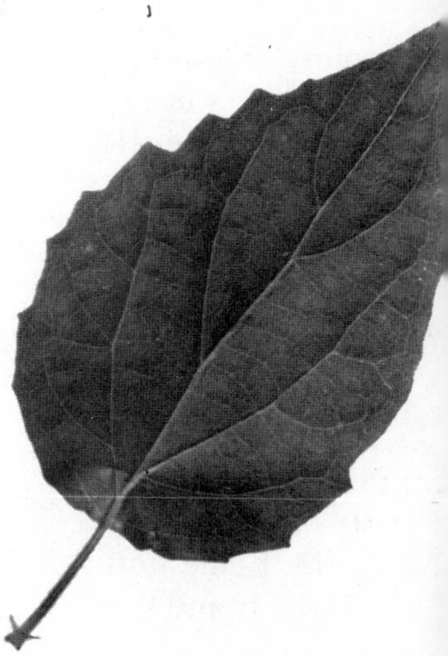

Maple-leaved Viburnum

Note leaves with irregular teeth, or wavy-edged.

← wavy-edged leaf-stalk

← reddish leaf-stalk

Nanny-berry

Black-haw

Withe-rod

Arrow-wood

Southern Arrow-wood

Smooth Withe-rod

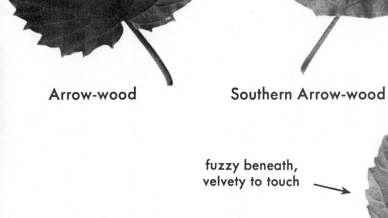

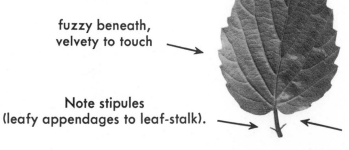

fuzzy beneath, velvety to touch →

Note stipules (leafy appendages to leaf-stalk). →

← short leaf-stalks (especially those of outermost pair of leaves on any twig)

Downy Viburnum

MP 139

naked buds (no scales
covering buds)

buds usually with six scales
(sometimes four)

brown buds

buds with four scales

flower
bud

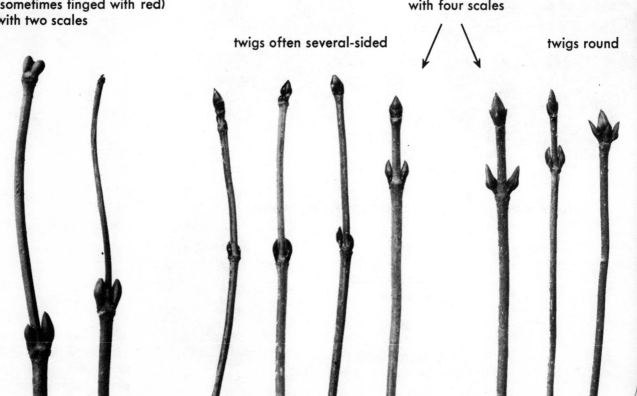

Hobble-bush

Downy Viburnum

Southern Arrow-wood

greenish buds
(sometimes tinged with red)
with two scales

brown buds
with four scales

twigs often several-sided

twigs round

Highbush Cranberry

Arrow-wood

Maple-leaved Viburnum

The buds on this page all have two scales.

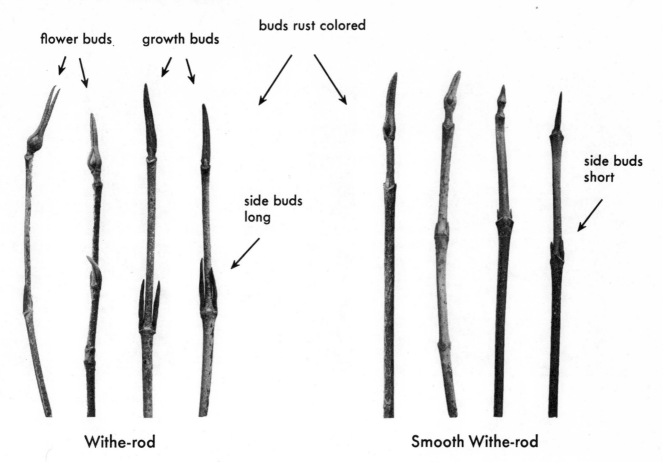

flower buds   growth buds

buds rust colored

side buds long

side buds short

Withe-rod

Smooth Withe-rod

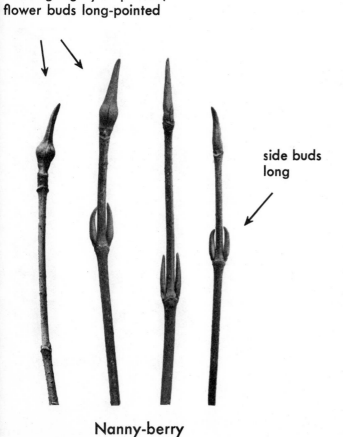

buds light gray or pinkish;
flower buds long-pointed

side buds long

Nanny-berry

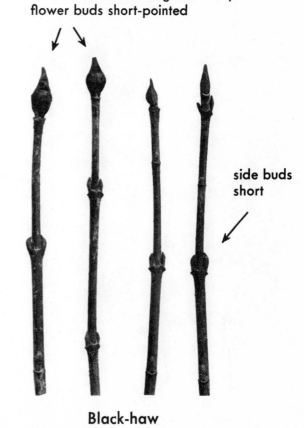

buds usually reddish
   (sometimes with slight bloom);
flower buds short-pointed

side buds short

Black-haw

MP 141

307

Nanny-berry

Hobble-bush

Black-haw

Highbush Cranberry

Arrow-wood

Downy Viburnum

Withe-rod

(Smooth Withe-rod and
Southern Arrow-wood similar)

Maple-leaved Viburnum

MP
143

BUTTONBUSH—*Cephalanthus occidentalis*

(Honeyballs; Globeflower)

Perfect July-Sept. Opposite usually in swamps or edges of ponds or streams; N.B. to Ont. & Minn., s. to Fla., Okla. & Tex., w. to Mex. & Calif.

15 ft.

leaves and buds opposite, frequently in whorls of threes or fours around the stems

ORALBERRY and SNOWBERRY—*Symphoricarpos* Perfect Opposite

Coralberry   S. *orbiculatus*   July-Aug.   Pa. to Minn., S. Dak. & Colo., s. to Fla. & Tex.;
   (Indian Currant; Buckbush)   as an escape, n. to N.E. & N.Y.   6 ft.
Snowberry   S. *albus*   May-Sept.   N.S. to Alb., s. to w. Mass., n. N.Y., Mich. &
   Minn.; as an escape, s. to Va.   4 ft.

flowers pinkish purple

fruit maroon or pinkish purple

flowers pink

fruit white

stems with pith   Coralberry   stems hollow   Snowberry

HONEYSUCKLE—*Lonicera* Perfect Opposite

I. VINES: all have hollow stems

Hairy Honeysuckle *Lonicera hirsuta*	May-July	Que. to Sask., s. to w. Vt., N.Y., Pa., Ohio, Mich., Minn. & Neb.; high-climbing
Japanese Honeysuckle *L. japonica*	June-Sept.	Asia; commonly escaped; semi-evergreen; trailing or high-climbing
Limber Honeysuckle *L. dioica* (Glaucous H.; Smooth-leaved H.; Mountain H.)	May-June	Me. & Que. to Sask., s. to Ga., e. Tenn., n.e. Ill. & Ia.; in var. to Kans.; bushy or slightly climbing
Trumpet Honeysuckle *L. sempervirens*	May-Sept.	Me. to Mich. & Ia., s. to Fla. & Tex.; semi-evergreen; high-climbing

II. UPRIGHT SHRUBS (MP 150)

  A. Stems with solid white pith; native

American Fly-Honeysuckle *L. canadensis* (Twinberry)	April-May	Me. & Que. to Sask., s. to s. N.E., Pa., Tenn. to Minn. 5 ft.
Mountain Fly-Honeysuckle *L. villosa* (Blue Honeysuckle)	May-July	Lab. to Man., s. to n. N.E., Pa., Mich. and Minn. Low (rarely 3 ft.)
Swamp Fly-Honeysuckle *L. oblongifolia*	May-June	N.B. to Man., s. to n. N.E., N.Y., Pa., Mich. & Minn. 5 ft.

  B. Hollow stems; foreign but escaping

Amur Honeysuckle *L. maackii*	May-June	e. Asia	15 ft.
European Fly-Honeysuckle *L. xylosteum*	May-June	Europe	10 ft.
Morrow Honeysuckle *L. morrowii*	May-June	Eurasia	8 ft.
Ta(r)tarian Honeysuckle *L. tatarica*	May-June	Eurasia	10 ft.

MP
146

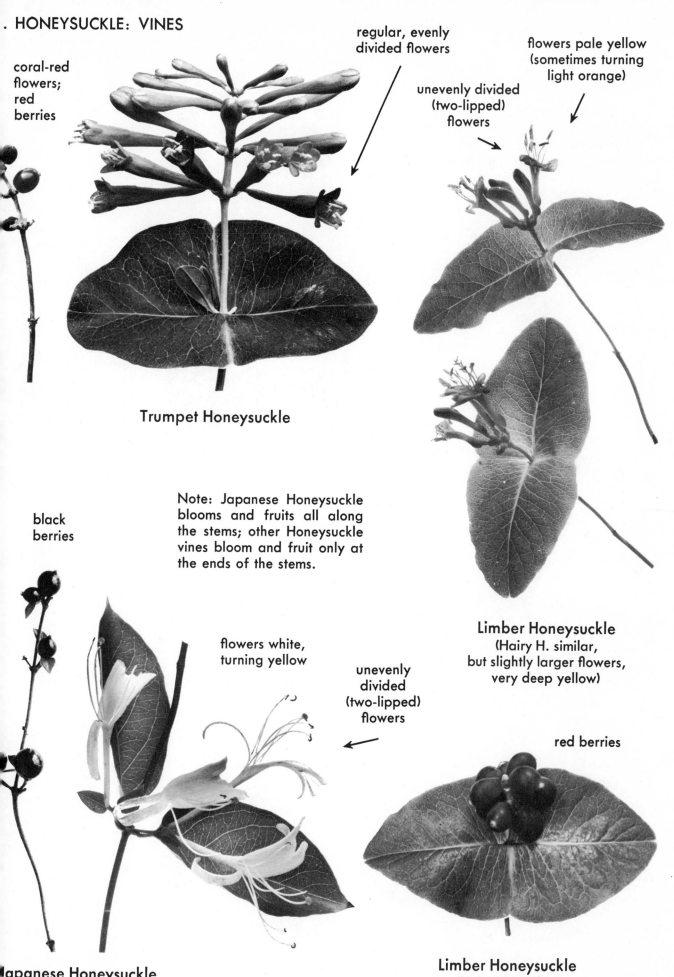

regular, evenly divided flowers

flowers pale yellow (sometimes turning light orange)

unevenly divided (two-lipped) flowers

coral-red flowers; red berries

Trumpet Honeysuckle

black berries

Note: Japanese Honeysuckle blooms and fruits all along the stems; other Honeysuckle vines bloom and fruit only at the ends of the stems.

Limber Honeysuckle
(Hairy H. similar, but slightly larger flowers, very deep yellow)

flowers white, turning yellow

unevenly divided (two-lipped) flowers

red berries

Japanese Honeysuckle

Limber Honeysuckle
(Hairy H. similar)

MP 147

I. HONEYSUCKLE: VINES (con't.)

The Honeysuckles on this page are semi-evergreen; those on the next page are deciduous.

Note connate leaf (two leaves joined together). The upper pairs of leaves of most Honeysuckle vines are connate. The Japanese Honeysuckle (shown at right) is one of the rare exceptions.

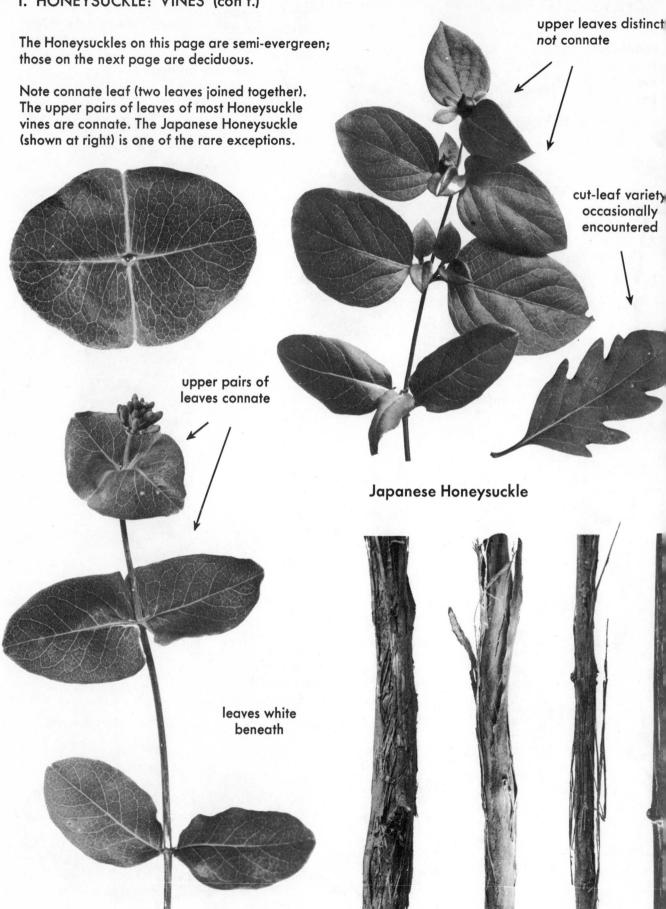

upper leaves distinct *not* connate

cut-leaf variety occasionally encountered

upper pairs of leaves connate

Japanese Honeysuckle

leaves white beneath

Trumpet Honeysuckle
(high-climbing)

barks typical of Honeysuckle vines
(all have hollow stems)

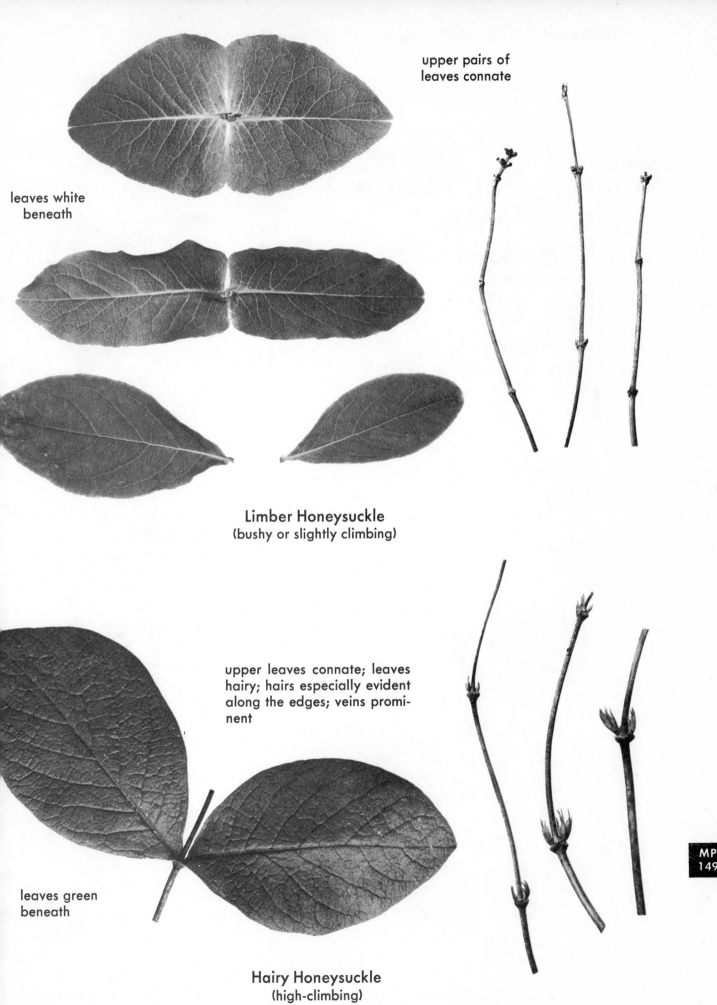

upper pairs of
leaves connate

leaves white
beneath

Limber Honeysuckle
(bushy or slightly climbing)

upper leaves connate; leaves
hairy; hairs especially evident
along the edges; veins prominent

leaves green
beneath

Hairy Honeysuckle
(high-climbing)

MP
149

II. HONEYSUCKLE: UPRIGHT SHRUBS   A. Solid Stems

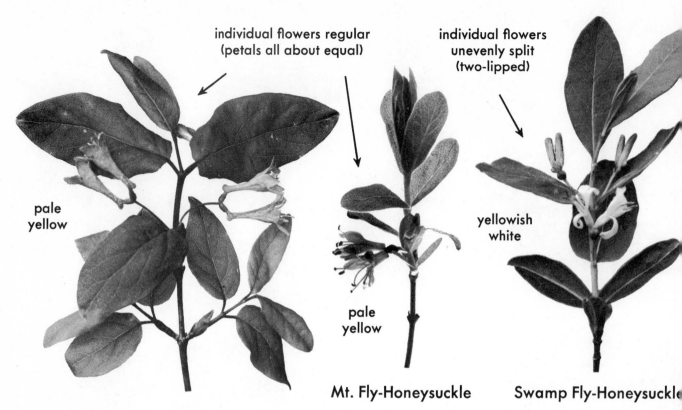

individual flowers regular (petals all about equal)

individual flowers unevenly split (two-lipped)

pale yellow

pale yellow

yellowish white

American Fly-Honeysuckle

Mt. Fly-Honeysuckle

Swamp Fly-Honeysuckle

B. Hollow Stems

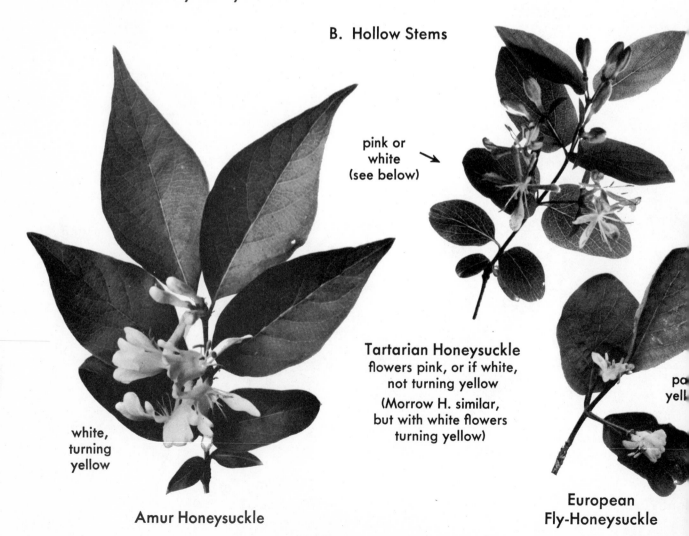

pink or white (see below)

white, turning yellow

Tartarian Honeysuckle
flowers pink, or if white, not turning yellow

(Morrow H. similar, but with white flowers turning yellow)

pale yell

Amur Honeysuckle

European Fly-Honeysuckle

A. Solid Stems

red

blue with bloom

red

American Fly-Honeysuckle

Mountain Fly-Honeysuckle

Swamp Fly-Honeysuckle

B. Hollow Stems

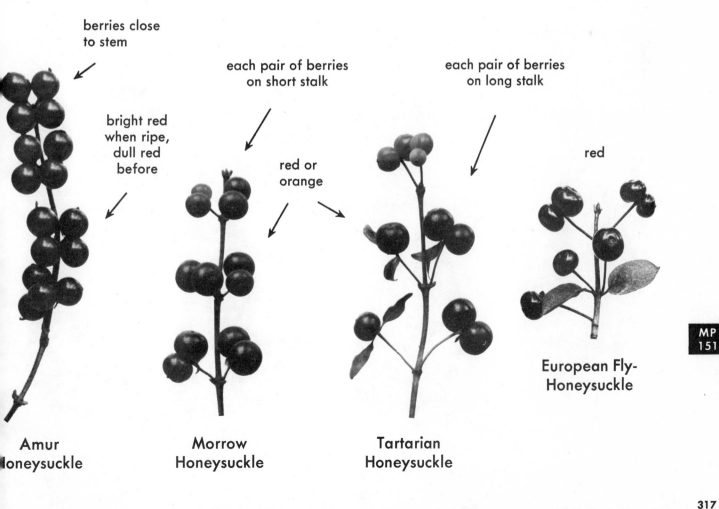

berries close
to stem

each pair of berries
on short stalk

each pair of berries
on long stalk

bright red
when ripe,
dull red
before

red or
orange

red

Amur
Honeysuckle

Morrow
Honeysuckle

Tartarian
Honeysuckle

European Fly-
Honeysuckle

MP
151

Note reddish leaf-stalks.

American Fly-Honeysuckle

Mountain Fly-Honeysuckle

Swamp Fly-Honeysuc

B. Hollow Stems

Amur Honeysuckle

Tartarian Honeysuckle
(Morrow H. similar)

European Fly-
Honeysuckle

MP
152

A. Solid Stems

B. Hollow Stems

American Fly-Honeysuckle

Mountain Fly-Honeysuckle

Swamp Fly-Honeysuckle

Amur, Morrow and Tartarian Honeysuckles similar

European Fly-Honeysuckle

typical Honeysuckle barks (both solid and hollow stems)

barks typical of all Honeysuckles included here (except Amur, shown at right)

Amur Honeysuckle

BUSH-HONEYSUCKLE — *Diervilla lonicera* Perfect May-July Opposite

Nfd. to Sask., s. to N.C., e. Tenn., n.e. Ill. & n.e. Ia.   3 ft.

Note: This is not a true Honeysuckle; although the flowers, bark and twigs are somewhat similar, the fruit of true Honeysuckles is a berry and their leaves do not have teeth (see MP 150-153).

yellow flowers

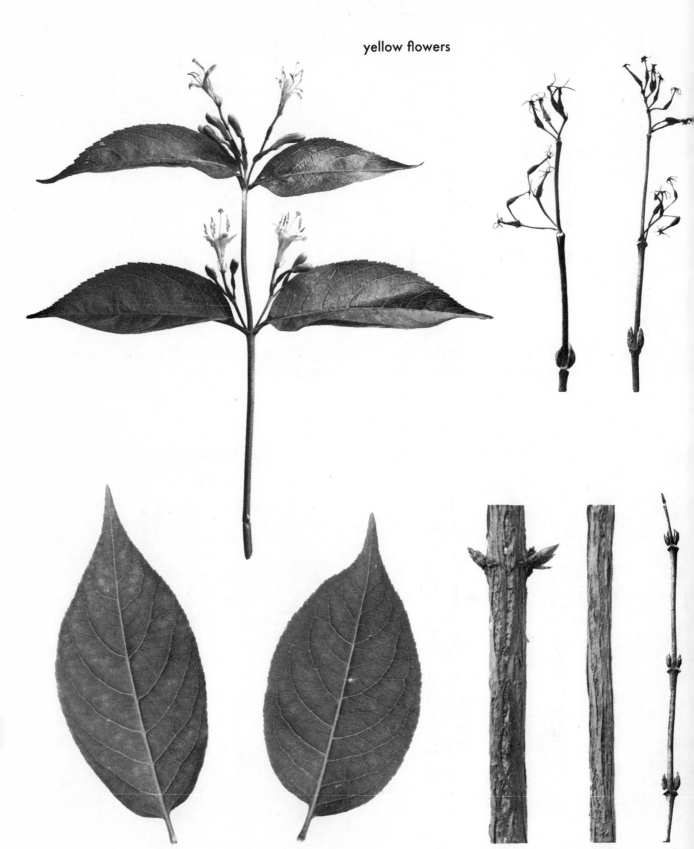

GROUNDSEL-BUSH — *Baccharis halimifolia*
(Consumption-weed; Sea Myrtle)

Dioecious   Aug.-Sept.   Alternate
coastal marshes, Mass. to Fla., Tex. & Mex.
12 ft.

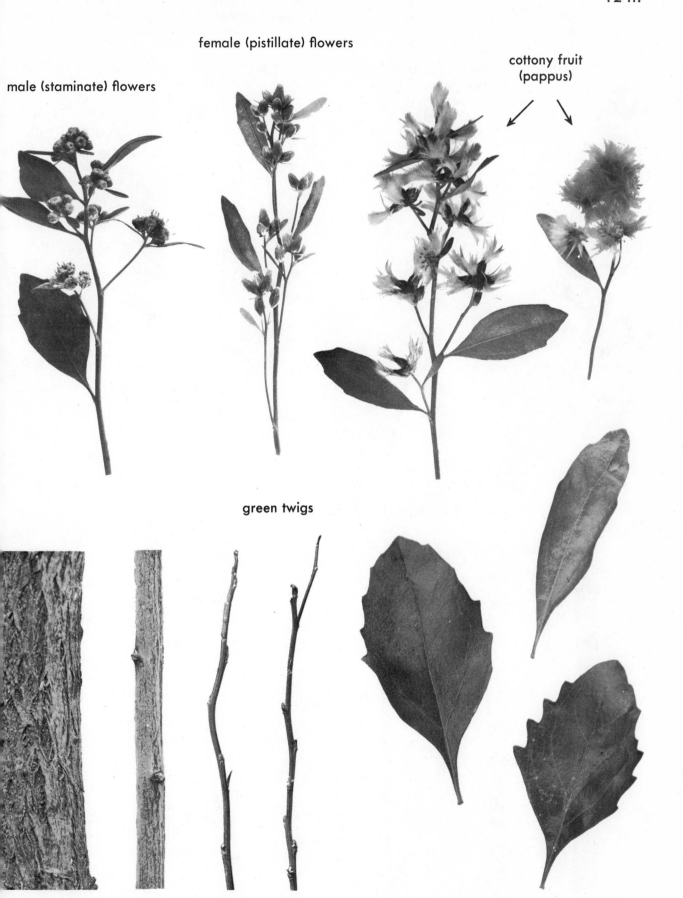

male (staminate) flowers

female (pistillate) flowers

cottony fruit
(pappus)

green twigs

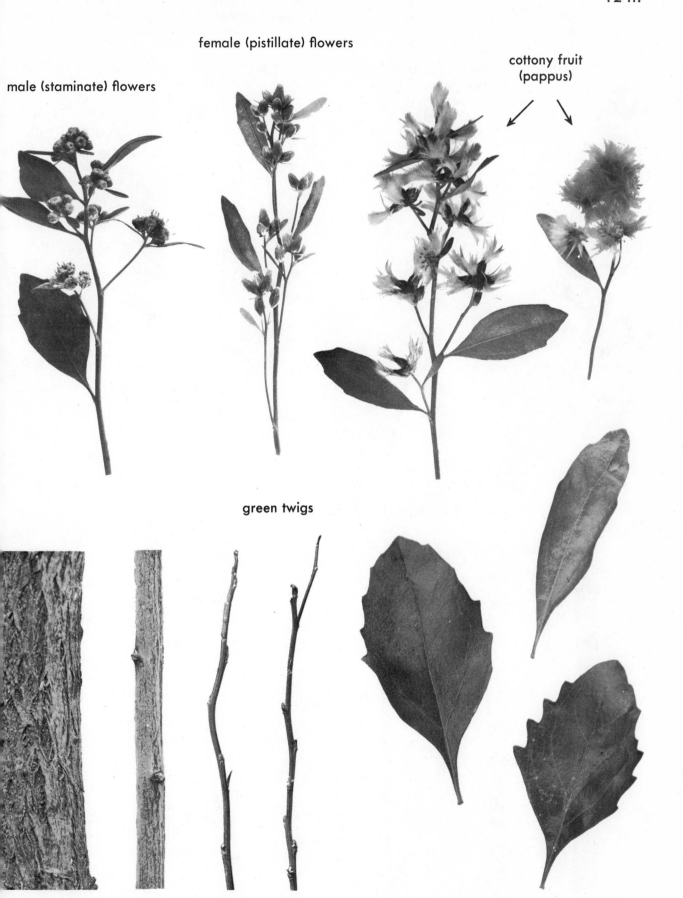

MP
155

MARSH-ELDER — *Iva frutescens oraria*   (see flower note below) Aug.-Sept.
mostly Opposite

The species *Iva frutescens* is a southern plant: Va. to Fla. & Tex. The variety given here is found in salt marshes from s. N.H. to Va. It is a subshrub growing to 3 ft. but dying back in winter.

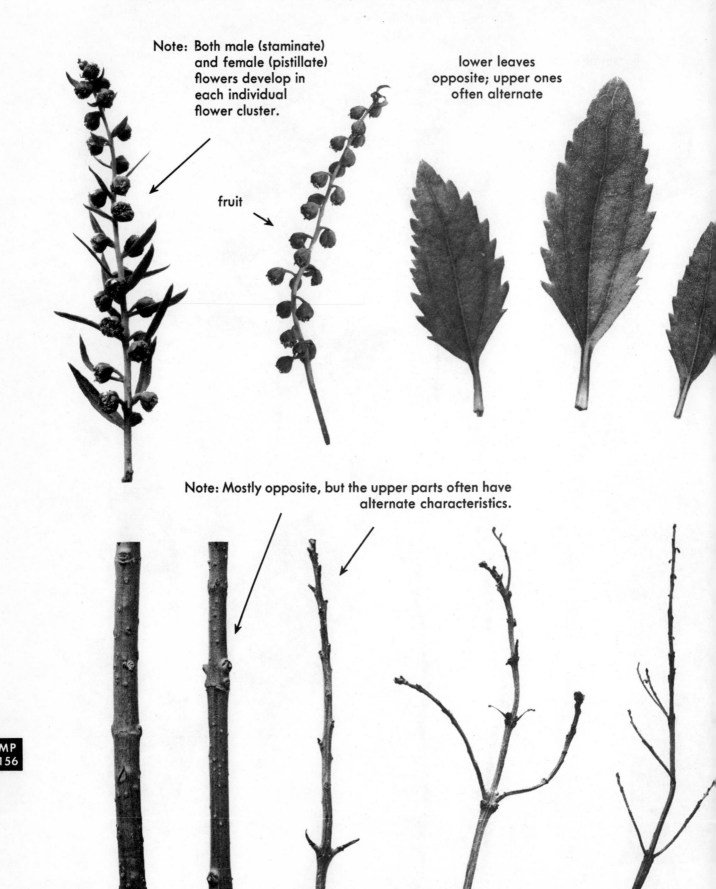

Note: Both male (staminate) and female (pistillate) flowers develop in each individual flower cluster.

lower leaves opposite; upper ones often alternate

fruit

Note: Mostly opposite, but the upper parts often have alternate characteristics.

NEEDLE-LEAVED SHRUBS
(all evergreen)

NEEDLE-LEAVED SHRUBS — *all evergreen*

The plants in this section include all Needle-leaved Shrubs, whether large or small, upright or prostrate. They comprise only ten genera and are shown on the following four pages. As there are so few plants, no Key is necessary, and both genus and species identification is accomplished by turning directly to this section whenever a Needle-leaved Shrub is in question. A quick perusal will indicate what is meant by the term "needle-leaved," and with the exception of the three plants below, which actually are not needle-leaved, there should be no confusion concerning which plants belong in this section.

The shrubs below will be found on the pages indicated for the Broad-leaved Upright Shrub section. Bog Rosemary, a close relative of Laurel, has very narrow leaves but they are not needle-like. Gorse produces no actual leaves, only leaf-stalks that are extremely sharp and thorn-like. Scotch Broom, although normally producing small tri-foliate leaves, sometimes is found with almost bare stems in summer, or with only a few undeveloped leaves as shown below.

MP
158
NE

Bog Rosemary 113A   Gorse 67B   Scotch Broom 65B

AMERICAN YEW—*Taxus canadensis*
(Ground Hemlock; Canada Yew)

Dioecious, rarely Monoecious   March
Nfd. to Man., s. to N.E., w. Va., W. Va.,
n. Ky., n. Ill., n.e. Ia. & Minn.   3 (occ. 6) ft.

inconspicuous flowers   conspicuous red fruit

Note: Leaves are yellow green beneath. (Hemlock, a tree, has somewhat similar-appearing needles, but they are whitish beneath.)

JUNIPER—*Juniperus*

Dioecious or Monoecious   Flowers inconspicuous

Common Juniper   *Juniperus communis*
   (Ground Juniper—*J. c. depressa*—is prostrate)

Creeping Juniper   *J. horizontalis*
   (Creeping Savin)

Europe & Asia; N. Amer.: s. to mts. of Ga., w. to Ill., Ia., & Calif.   6 ft. (rarely a tree, to 35 ft.)

N.S. to Alb., s. to n. N.E., n. N.Y., Mich. & west
                                    Prostrate

fruit blue with bloom

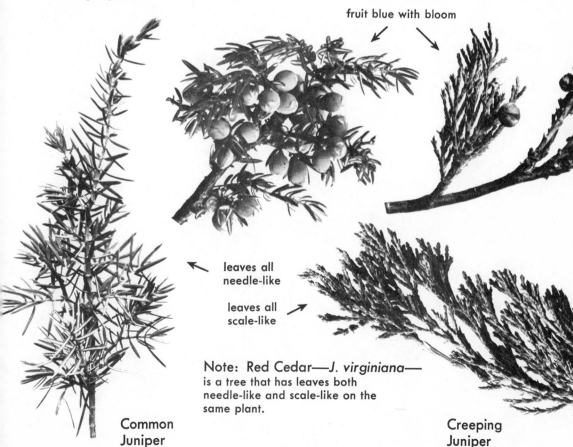

← leaves all needle-like

leaves all scale-like →

Note: Red Cedar—*J. virginiana*—is a tree that has leaves both needle-like and scale-like on the same plant.

Common Juniper

Creeping Juniper

BROOM CROWBERRY—*Corema conradii*

Dioecious   April-May
rare: Nfd. near coast to Mass., Shawangunk Mts. of
N.Y., & N.J. pine barrens   18-24 in.

male (staminate)
flowers purple
brown to brown

female (pistillate)
flowers brown

fruit minute and
inconspicuous

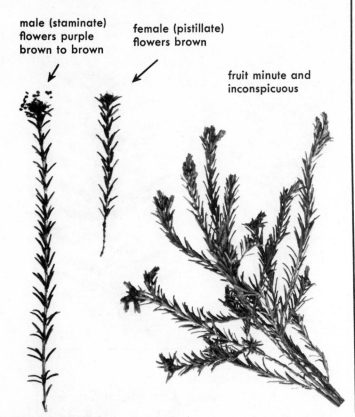

CROWBERRY—*Empetrum nigrum*

(Curleberry)   Dioecious or Monoecious   April-May
circumpolar, s. to Me. coast, alpine areas of N.E., N.Y.,
one isolated spot on e. L.I., n. Mich., n. Minn. & n. Calif.
                                              Prostrate

flowers inconspicuous,
along stems and lost
in among the leaves

fruit black
or purple

HEATH—*Erica*   Perfect   Europe; naturalized e. U.S.

There are a number of species, two of the most common
being:

Cornish Heath   *Erica vagans*   July-Oct.   1½-2 ft.
Spring Heath   *E. carnea*   Feb.-May   1½-2 ft.

Most Heaths are late-bloomers,
but Spring Heath blooms in late
winter and early spring. Heath
flowers vary from rosy red or
pinkish lavender to white, rarely
yellow.

HEATHER—*Calluna vulgaris*

Perfect   July-Sept.
Europe, e. Asia; naturalized e. U.S.
1½ (rarely 3) ft.

flowers lavender
to pink

MOUNTAIN HEATH—*Phyllodoce coerulea*
Perfect   June-Aug.
circumpolar, s. to alpine areas of Me. & N.H.   6 in.

flowers light lavender
to purple

CASSIOPE—*Cassiope hypnoides*   Perfect   June-July
circumpolar, s. to alpine areas of Me. & N.H. Prostrate; tufts

flowers white
on red stems
with red calyx
(star at base
of flower)

BEACH HEATHER—*Hudsonia*   Perfect
Golden Heather   *H. ericoides*   May-June   Nfd. & N.B., s. to N.H. & interruptedly s. to N.J., Va. & N.C.   12 in.
Woolly Heather   *H. tomentosa*   May-June   chiefly near coast, N.B. to N.C., also near Great Lakes; in yar. farther
                                              inland   12 in.

Note: The individual needles of Woolly Heather are much closer to the stems and less clear-cut than those of Golden Heather. Flowers and fruit of Woolly Heather have little or no stalks (sessile), whereas those of Golden Heather are on distinct stalks.

Both Beach Heathers have yellow flowers.

Note distinct
flower stalks.

Note that fruit is on
distinct stalks.

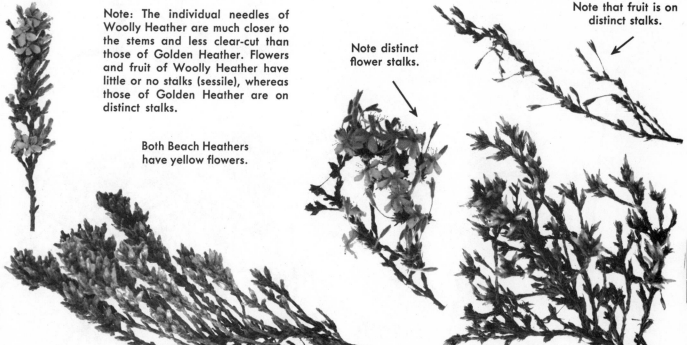

Woolly Heather

Golden Heather

CLUBMOSS—*Lycopodium*
(Ground Pine)

This genus does not have true flowers; reproduction is chiefly by means of spores. The Clubmosses are found in most of the range covered here.

Ground Cedar—*Lycopodium complanatum*   fruits July-Sept.
(Creeping Jenny; Running or Trailing Evergreen; Ground Pine)

Note: Another Clubmoss, *L. tristachyum*, has many of the same common names as *L. complanatum*, and except for being more deeply rooted, is almost identical with it.

Running Clubmoss—*L. clavatum*   fruits July-Sept.
(Coral or Staghorn Evergreen; Buckhorn; Wolf's-claws)

Shining Clubmoss—*L. lucidulum*   fruits July-Sept.

Note: The fruiting spores are along the stems and not terminal, as with the other Clubmosses shown here.

Tree Clubmoss—*L. obscurum*   fruits July-Nov.
(Flat-branch Ground Pine)

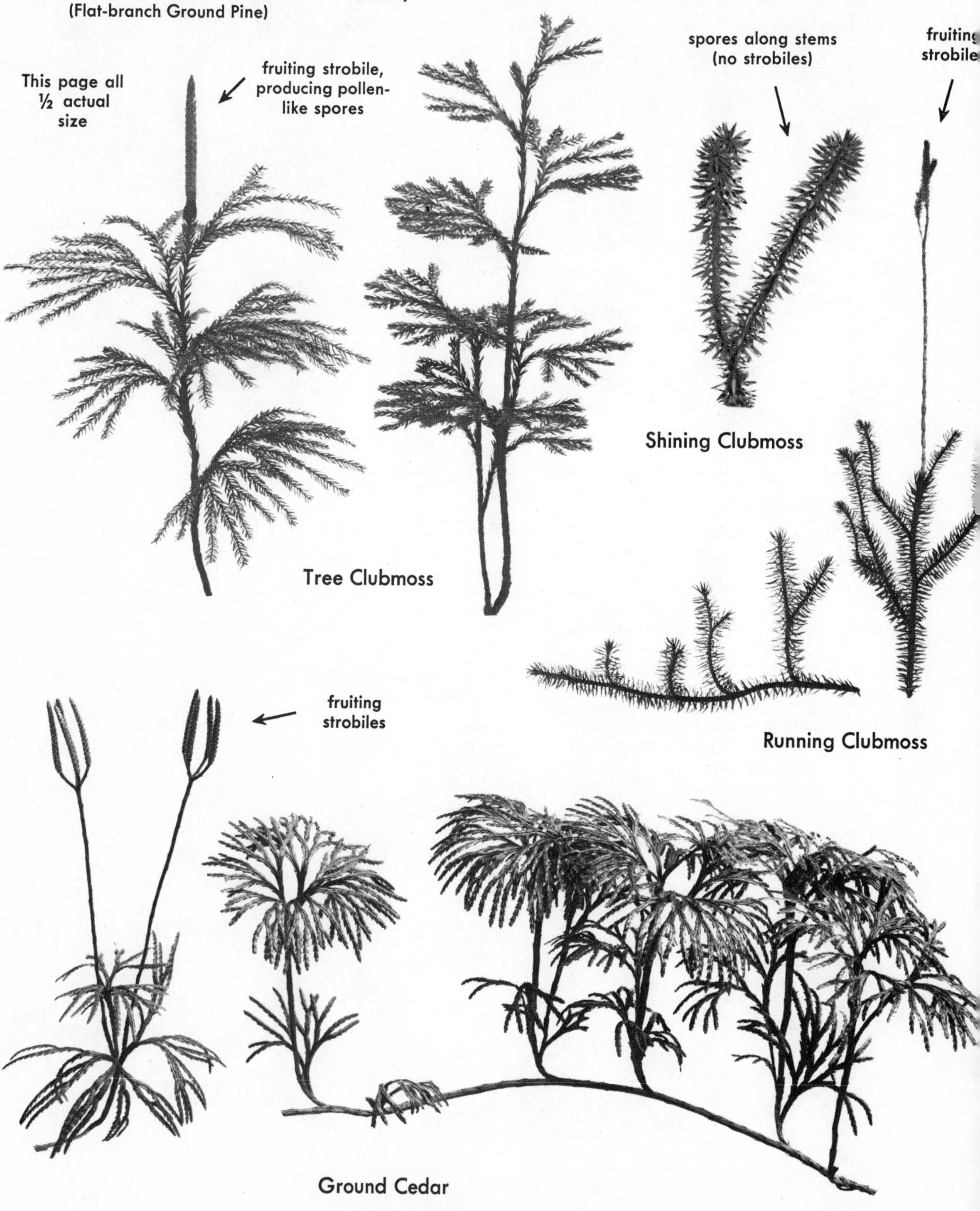

spores along stems
(no strobiles)

fruiting
strobile

This page all
½ actual
size

fruiting strobile,
producing pollen-
like spores

Shining Clubmoss

Tree Clubmoss

Running Clubmoss

fruiting
strobiles

Ground Cedar

BROAD-LEAVED GROUND COVERS

The plants shown on the following eight pages are all small, broad-leaved plants that normally do not grow over 12 inches in height (mostly less). Some of them cover a considerable area of ground, forming dense mats. Some are tufted small plants; others have fairly long runners, but do not have heavy woody stems. The distinction between a trailing ground cover and a trailing vine is purely arbitrary, and size has been the deciding factor in this case. A perusal of both sections will make this point clear.

Some ground covers belong to genera that include upright shrubs. In such a case, a cross reference is made to the Master Page giving both types together. In all other cases, the ground covers are shown here only.

Sand Myrtle, below, is not a ground cover, but a small upright shrub; it is shown here as it is sometimes quite low and might be confused with one of the small evergreen ground covers. It will be found in the Upright Shrub section, as indicated below.

Sand Myrtle 113B

MP
163

GC

COMMON BEARBERRY—*Arctostaphylos uva-ursi*
(Kinnikinic; Mealberry)

Perfect   April-June   mostly Alternate
N. Amer.: s. to Va., Ill. & west
              Prostrate; creeping

flowers white,
tinged with pink

fruit red

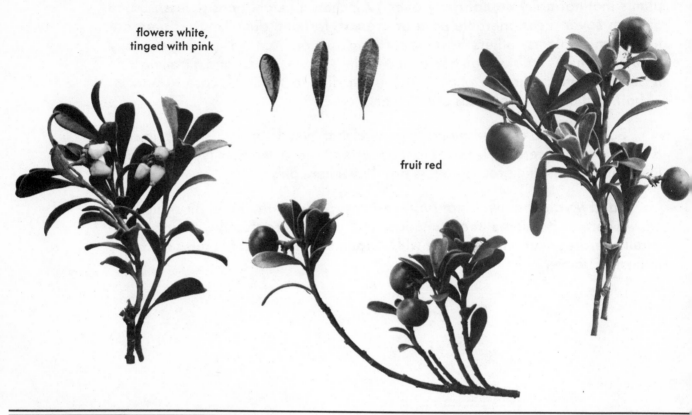

PACHISTIMA—*Pachistima canbyi*
(Mountain Lover; Cliff Green; Rat Stripper)

Perfect   April-May   Opposite
mts. of Va. and W. Va.; sparingly in Ky. & Ohio
              Matted, 10-12 in.

flowers greenish brown

fruit an
inconspicuous
capsule

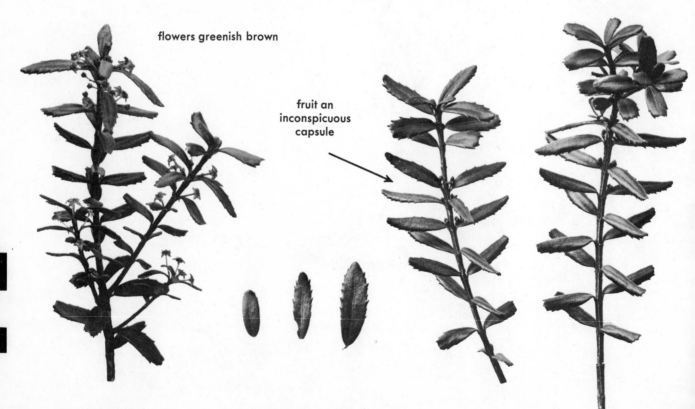

MP
164

GC

330

CRANBERRY—*Vaccinium*   (see MP 119)   Perfect   Alternate

Large Cranberry—*V. macrocarpum*	June-Aug.	Nfd. to Sask., s. to N.C., Tenn., Ill., Wis. & Minn. Creeping
Small Cranberry—*V. oxycoccus*	June-July	N. Amer.: s. to N.E., N.J., W. Va., Ill., Wis., Minn. & west Creeping

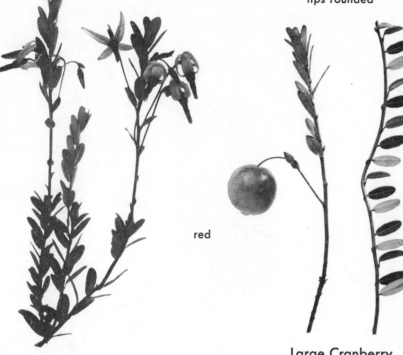

leaves largely flat;
tips rounded

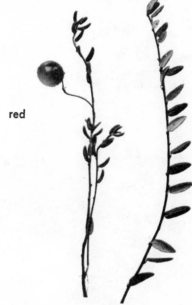

leaves curl over at
edges; tips pointed

red

red

Large Cranberry   **Small Cranberry**

EVERGREEN HUCKLEBERRY

(Box-Huckleberry)   *Gaylussacia brachycera*
(see MP 126)   Perfect April-June Del. & Pa. to Tenn.
Creeping, with ascending stems to 18 in.

flowers white, tinged
with pink and red

fruit blue
with bloom

FLOWERING MOSS—*Pxyidanthera barbulata*

(Pyxie)   Perfect   April-May
N.J. to N.C. in sandy barrens   Prostrate tufts

flowers white;
fruit inconspicuous

MP
165

GC

BEARBERRY WILLOW—*Salix uva-ursi*
(see MP 1)   Dioecious   June-July
Lab. to Alaska, s. to mts. of n. N.E. & n. N.Y.
Prostrate; matted

reddish female
(pistillate)
flowers

grayish or yellow
male (staminate)
flowers

ALPINE BEARBERRY—*Arctous alpinus*
Perfect   April-June
Arctic, s. to mts. of Me. & N.H.   Prostrate; matte

black or
purplish
berries

THREE-TOOTHED CINQUEFOIL
Potentilla tridentata   (see MP 64)
Perfect   May-Oct.   Arctic, s. to N.E. & N.Y., w. to
N. Dak., s. in mts. to Ga.; woody only at base
Low; matted

white
flowers

BILBERRY—*Vaccinium*   (see MP 119)   Perfect
Bog Bilberry   *V. uliginosum*   May-June   n. N. Amer., s
(Alpine B.)   to N.E., n. N.Y., Mich. & Minn.   Low; prostrate
Dwarf Bilberry   *V. cespitosum*   May   n. N. Amer. s. to
N.E., n. N.Y., Mich., Minn. & west   Low; tufts

leaves thin,
green

white to
deep pink

Dwarf Bilberry

typical
Bilberry flowers

blue
or purple
with bloom

leaves
leathery;
bluish green

Bog Bilberry

typical
Bilberry fruit

LAPLAND RHODODENDRON
Rhododendron lapponicum (see MP 104)
Perfect June-July Arctic, s. to mts. of Me. N.H. & N.Y.; also in Wis. Prostrate; matted

flowers purple or magenta

leaves pitted above, many small brown spots (scales) beneath

MOUNTAIN CRANBERRY
Vaccinium vitis-idaea minus (see MP 119)
Perfect May-June Arctic, s. to N.E., n. Minn. & west Matted, 6-12 in.; prostrate at higher elevations

flowers white, tinged with pink

berries red, usually small at high altitudes, larger elsewhere

ALPINE AZALEA—*Loiseleuria procumbens*
Perfect June-Aug. Arctic, s. to mts. of Me. & N.H. Prostrate; matted

leaves with definite groove on upper side

fruit red at first, brown when mature

pink flowers

DIAPENSIA—*Diapensia lapponica*
Perfect June-July Arctic, s. to mts. of N.H. & N.Y. Prostrate; tufts

fruit

leaves without definite groove on upper side

white flowers with very yellow stamens

MP 167

GC

PARTRIDGE-BERRY—*Mitchella repens*   Perfect   June-Aug.   Opposite
N.S. to Ont. & Minn., s. to Fla., Ark. & Tex.   Trailing

flowers fuzzy inside

red berries

TWINFLOWER—*Linnaea borealis americana*
Perfect   June-Aug.   Opposite
Lab. to Alaska, s. to L.I., W. Va., Wis. & west   Trailing

fruit
fuzzy

flowers pink,
red inside

CREEPING SNOWBERRY—*Chiogenes hispidula*
Perfect   May-June   Alternate
Nfd. to B.C., s. to N.C., W. Va., Wis., Minn. & west
Trailing

flowers
inconspicuous,
on underside
of stems

Leaves and fruit
have wintergreen
odor and taste
when crushed.

white
berries

MP
168

GC

BUNCHBERRY—*Cornus canadensis*
(Dwarf Cornel; Dwarf Dogwood)

(see MP 98)   Perfect   May–June   Deciduous   Opposite
(leaves in whorls at ends of stems)
Arctic, s. to Md., W. Va., Ill. & west   Upright, 6-10 in.

red
berries

TRAILING ARBUTUS—*Epigaea repens*
(Mayflower)

Perfect   April–May   Evergreen   Alternate
Lab. to Sask., s. to Ga., Tenn., Ind., Wis. & Minn.
Prostrate; trailing

flowers white,
tinged with
pink; very
fragrant

Covering opens,
when mature, to
expose seeds on
surface of fruit,
like a strawberry.

MP
169

GC

WINTERGREEN—*Gaultheria procumbens*
(Checkerberry; Teaberry)

Perfect   May-Sept.   Nfd. to Man., s. to Ga. & Ala. Upright plants to 6 in., with underground stems, often covering considerable area

Leaves and fruit have wintergreen odor and taste when crushed.

fruit red

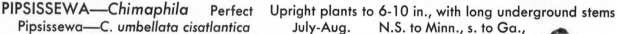

PIPSISSEWA—*Chimaphila*

Perfect   Upright plants to 6-10 in., with long underground stems

Pipsissewa—C. *umbellata cisatlantica*   July-Aug.   N.S. to Minn., s. to Ga., n.e. Ill. & Wis.
   (King's Cure; Wintergreen; Prince's Pine)

Striped Pipsissewa—C. *maculata*   June-Aug.   s. N.H. to s. Ont., s. Mich. & n.e. Ill., s. to Ga., Ala. & Tenn.
   (Spotted Wintergreen; Spotted Ratbane)

flowers typical of both species; white, sometimes pink

fruit typical of both species

Striped Pipsissewa

Pipsissewa

PACHYSANDRA—*Pachysandra terminalis*
(Japanese Spurge)

Monoecious   April-May

Japan; escaped elsewhere   Mat-forming; 10-12 in.

Note: Allegheny Spurge—*P. procumbens*—is a southern species, semi-evergreen, with less glossy leaves, producing flowers at the base of the stem instead of at the top, above the leaves, as seen here for Japanese Spurge.

Note: The top flowers of the cluster are male (staminate). The bottom ones (two visible here) are female (pistillate). The male flowers soon disintegrate, leaving only female ones which develop into the white berry-like fruit shown below.

fruit shown not fully developed (rounder, and soft, when mature)

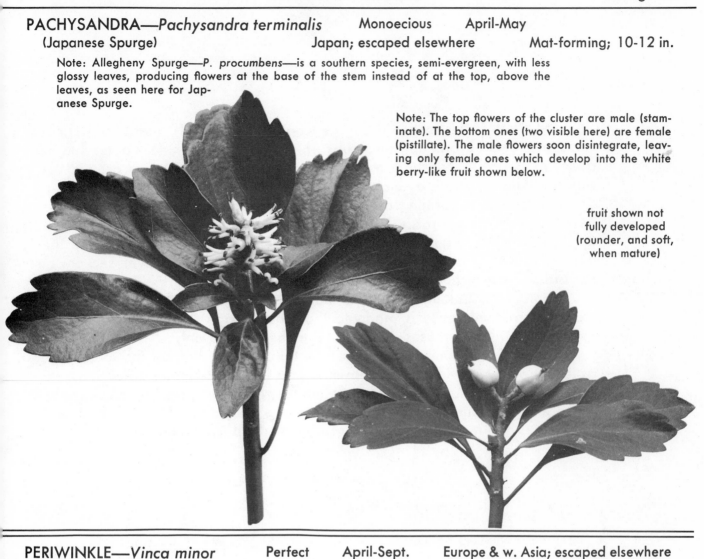

PERIWINKLE—*Vinca minor*
(Myrtle)

Perfect   April-Sept.

Europe & w. Asia; escaped elsewhere
Long creeping stems; mat-forming

flowers blue (white or purple in variety)

fruit

Note: There is also a form with variegated leaves.

MP
171

GC

VINES

The plants in this section are large woody plants which, under normal circumstances, never form upright shrubs. They may climb by means of tendrils or aerial rootlets attached to some type of support, or by twining around the support. If no support is available, the climbing plants can only trail along the ground, sometimes forming dense mats. Several plants that are not climbers never take an upright form, but trail along the ground or scramble over low objects. They are included here as they have long woody stems unlike the slender growth of somewhat similar plants in the Ground Cover section. A perusal of both sections will make the distinction clear.

Some vines belong to genera that also include upright shrubs. In such cases, a cross reference is made to the Master Page that gives both types together. In all other cases, the vines are shown here only.

Rambler Roses and the so-called Matrimony-vine are not included here, as they are not true vines. They sometimes scramble over other objects, but do not really climb. If no support is available, they become upright, arching shrubs and are never found prostrate or trailing along the ground. They will be found in the Keys and Master Pages for Broad-leaved Upright Shrubs.

ARRANGEMENT OF VINES

I Alternate Leaves and Buds

(1) Vines climbing by tendrils
Tendrils grow only at the nodes
(joints)
	MP 173-181	Ampelopsis
		Grape
		Woodbine
		  Boston Ivy
		  Virginia Creeper
		Green-Brier or Cat-Brier

(2) Vines climbing by aerial rootlets Rootlets are not confined to nodes, but appear along whole sections of stem	MP 182-183	English Ivy Poison Ivy
(3) Vines climbing by twining No tendrils or rootlets	MP 184-191	Moonseed Dutchman's Pipe Bittersweet Wisteria Nightshade
(4) Vines trailing or scrambling	MP 192-193	Currant Dewberry

II Opposite Leaves and Buds   MP 194-200   Clematis
    As there are only four genera, no sub-   Eunonymus
      divisions are given   Honeysuckle
                                        Trumpet-vine

HEARTLEAF AMPELOPSIS—*Ampelopsis cordata*

Perfect   May-June   High-climbing
Va., Ohio, Ill. & s.e. Neb., s. to Fla., Tex. & Mexico; naturalized n. to Mass.

Note: Pepper-vine or Thunderberry—*A. arborea*—is a southern vine with compound leaves.

Porcelainberry—*A. brevipedunculata*—from Asia, but occasionally escaped, has leaves with three to five lobes.

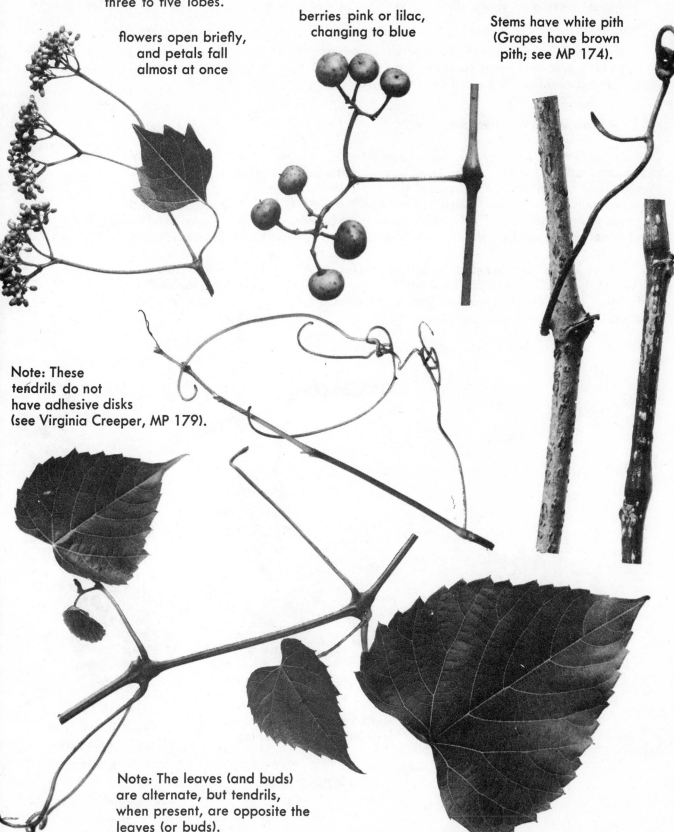

flowers open briefly, and petals fall almost at once

berries pink or lilac, changing to blue

Stems have white pith (Grapes have brown pith; see MP 174).

Note: These tendrils do not have adhesive disks (see Virginia Creeper, MP 179).

Note: The leaves (and buds) are alternate, but tendrils, when present, are opposite the leaves (or buds).

MP
173

V

GRAPE—*Vitis* Dioecious or Polygamous Mostly high-climbing

Fox Grape *V. labrusca* May-June N.E. to Mich., s. to Ga., Tenn. & Ill.
 (This is the parent of the Concord, Catawba, Chautauqua and many other cultivated grapes.)

Muscadine Grape *V. rotundifolia* June Del. to Fla., w. to Mo., Okla. & Tex.
 (Scuppernong)

Riverbank Grape *V. riparia* May-June N.S., N.B. & Man., s. to Va., Mo., Tex. & N. Mex.

Summer Grape *V. aestivalis* May-June Vt. to Mich. & Wis., s. to Ga. & Tex.

 (The Silverleaf Grape—*V. a. argentifolia*—a variety of the Summer Grape, extends its range to N.H. & Minn. Rehder considered the Silverleaf Grape to be a distinct species: *V. argentifolia*; but it is now usually treated as a variety as given here.)

 All Grapes have brown pith.

 Most Grapes are high-climbers, have shreddy, peeling bark, stems with pith separated at the nodes by a solid woody diaphragm. They usually have tendrils forking into two or more parts toward the ends. There is one marked exception to this, the Muscadine Grape, which has relatively smooth bark, continuous pith and tendrils that do not fork. (The Bush Grape—*V. rupestris*—has few, if any, tendrils and is not a high-climber, but in other respects it is similar to other Grapes.)

 There are a good number of Grape species, which tend to be confusing because of the small differences between them; in addition, considerable hybridization takes place and intermediate forms are common. For this reason, only a few main types are given here to represent the genus.

woody
diaphragm

typical
Grape flowers

Fox Grape

dark blue, black, red, brown
or green, usually with bloom

(Muscadine G. similar;
dull purple or bronze;
no bloom; very tough skin)

Summer Grape

dark blue to black,
usually with bloom

(Riverbank G. similar,
usually with heavier bloom)

MP
174

V

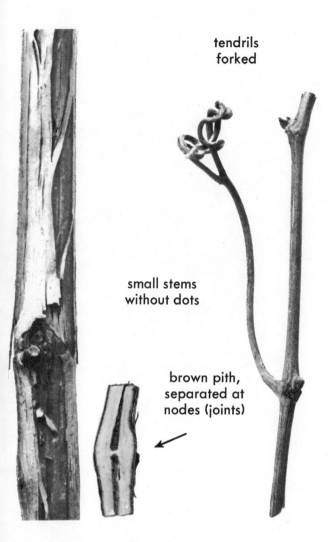

tendrils
forked

small stems
without dots

brown pith,
separated at
nodes (joints)

Summer Grape
(Fox G. and Riverbank G. similar,
but see distinctions below)

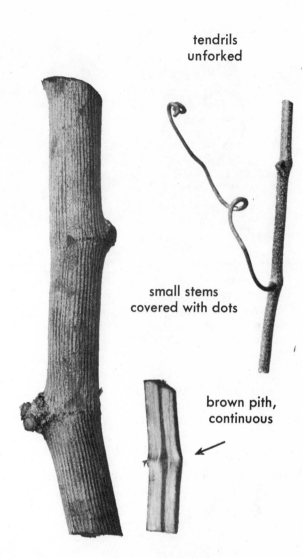

tendrils
unforked

small stems
covered with dots

brown pith,
continuous

Muscadine Grape

(1) **Fox Grape**
tendrils (or flowers) opposite
at least three consecutive
leaves (or buds)

(2) **Riverbank G. and Summer G.**
(also most other Grapes)
tendrils (or flowers) usually
missing opposite at least every
third leaf (or bud)

Note: The leaves and buds of
all Grapes are alternate, but
the tendrils are opposite the
leaves or buds.

MP ·
175

V

The leaves on this page are rusty or pale beneath.

early in season,
usually pale beneath;
later, very rusty beneath

Fox Grape

MP
176

V

pale at first;
later in season,
often somewhat
rusty beneath

Summer Grape
(Silverleaf G. similar, but very white and smooth beneath)

The leaves on this page are green beneath.

Riverbank Grape

MP
177

V

Muscadine Grape

WOODBINE—*Parthenocissus*   Perfect (rarely Polygamous)   High-climbing

Boston Ivy	*P. tricuspidata*	June-Aug.	Japan, China; locally escaped
Virginia Creeper	*P. quinquefolia*	June-Aug.	Me. & Que. to Minn., s. to Fla.,
(Woodbine; Five-leaved Ivy)			Tex. & Mexico

berries blue to black
with slight bloom,
borne on red stalks

Virginia Creeper

berries blue to black
with bloom

Boston Ivy

Note: Both Boston Ivy and Virginia Creeper usually have adhesive disks at the tips of the tendrils.

Note: Boston Ivy has similar bark; the stems of both species have white pith.

Virginia Creeper

mature (or adult) leaf produced on fruiting stems

juvenile leaf found on young plants and bottom shoots

Note: The leaves (and buds) of Boston Ivy and Virginia Creeper are alternate, but the tendrils are opposite the leaves (or buds).

MP 179

V

Boston Ivy

GREEN-BRIER or CAT-BRIER — *Smilax*   Dioecious   High-climbing

Bristly (or Hispid) Green-Brier	*S. hispida*	June	Conn. to Ont., Minn. & S. Dak., s. to Ga. & Tex.
Cat-Brier (Saw-Brier; Glaucous Green-Brier)	*S. glauca*	June	Mass. & s.e. N.Y. to Mo., s. to Fla. & Tex.
Horse-Brier (Common G.; Bull-Brier; Round-leaved Brier)	*S. rotundifolia*	May-June	N.S. near coast to s.e. N.H., w. from Mass. to N.Y., s. Ont. & s. Mich., s. to Fla. & Tex.
Laurel-leaved Green-Brier	*S. laurifolia*	July-Sept.	N.J., Va. & Tenn., s. to Fla. & Tex., evergreen
Red-berried Green-Brier	*S. walteri*	June	N.J., Va. & Tenn., s. to Fla. & La.

typical Green-Brier flowers

female (pistillate)

male (staminate)

berries blue to black, sometimes with bloom

Horse-Brier

similar:
 Bristly G.
 Cat-Brier
 Laurel-leaved G.
similar, but with
 red berries: Red-
 berried G.

evergreen

Laurel-leaved Green-Brier

green beneath

whitish beneath when fully developed

Bristly Green-Brier

Horse-Brier
(Red-berried G. similar)

Cat-Brier

MP
180

V

346

thorns and bristles distinctly blackish

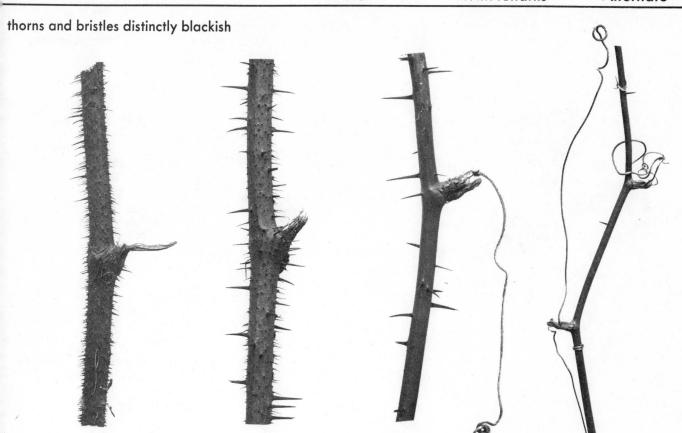

Bristly Green-Brier

thorns mostly green

thorns
at nodes
(joints)
as well as
in between

no thorns
at nodes
(joints)

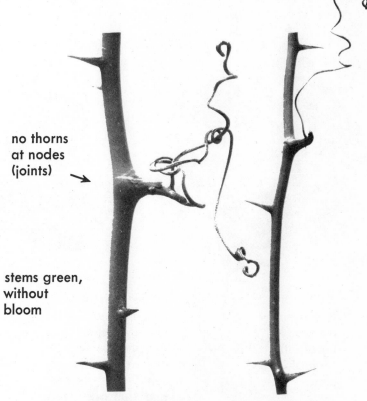

stems green,
usually with
some bloom

stems green,
without
bloom

Cat-Brier

Horse-Brier (thorns along entire length)
(Laurel-leaved G. and Red-berried G. similar, except that they produce thorns only on the lower
parts of the stems)

ENGLISH IVY — *Hedera helix*   Perfect   Aug.-Oct.   Evergreen
Europe & e. Asia; escaped   High-climbing

dark blue
to black

Note aerial rootlets
used in climbing.

leaf from
mature
(or adult)
form of
plant

Note: English Ivy
may not mature for
many years, and
the juvenile form is
the one commonly
seen. The flowers
and fruit are rarely
seen, as only the
adult form can pro-
duce them.

leaves from
juvenile form
of plant

MP
182

V

POISON IVY — *Rhus radicans*   (see MP 74: Sumac)   Dioecious   June-July
N.S. to Minn., s. to Fla. & Ariz.   High-climbing

male (staminate) flowers
whitish

female (pistillate) flowers
greenish

fruit dull
white

aerial rootlets not always
evident when not climbing

climbs by means
of aerial rootlets

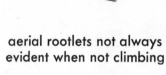

MP
183

V

349

COMMON MOONSEED — *Menispermum canadense*   Dioecious   May-June
w. Que. & w. N.E. to Man., s. to Ga. & Okla.   High-climbing

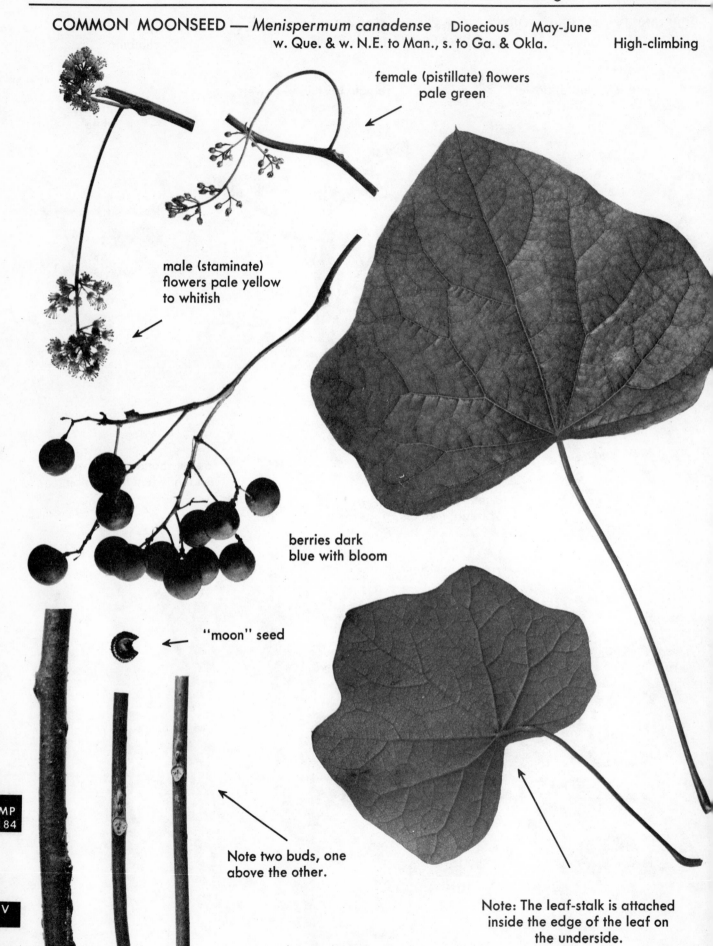

female (pistillate) flowers
pale green

male (staminate)
flowers pale yellow
to whitish

berries dark
blue with bloom

"moon" seed

Note two buds, one
above the other.

Note: The leaf-stalk is attached
inside the edge of the leaf on
the underside.

MP
184

V

350

DUTCHMAN'S PIPE — *Aristolochia durior*

Perfect   May   w. Pa., s. to Ga. & Ala.;
widely planted & naturalized outside this area   High-climbing

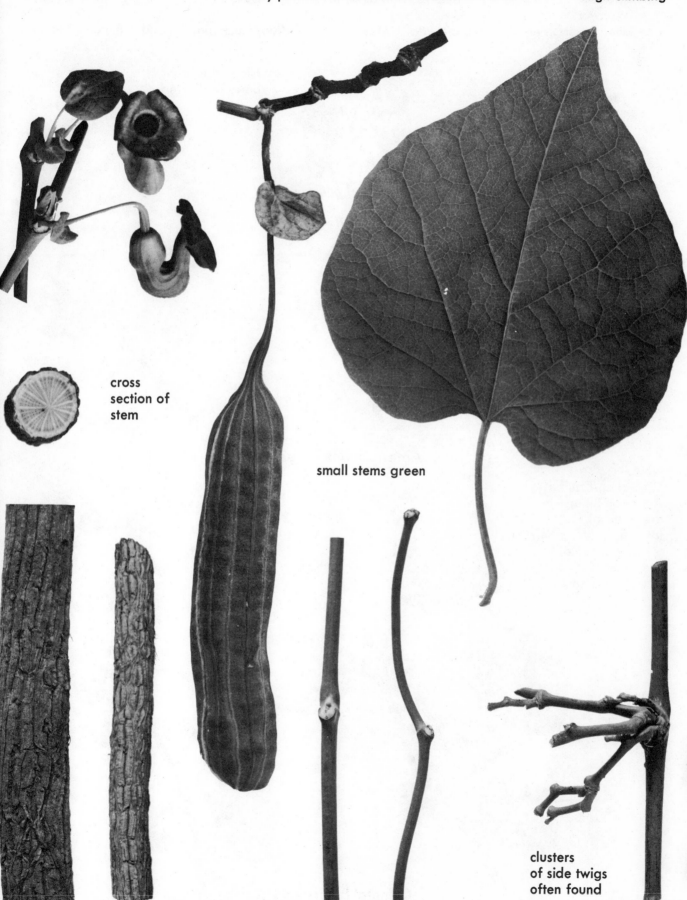

cross
section of
stem

small stems green

clusters
of side twigs
often found

MP
185

V

BITTERSWEET — *Celastrus*   Dioecious (rarely Polygamous)   High-climbing

American Bittersweet (Waxwork)	*C. scandens*	May-June	Me. & Que. to Man., s. to Ga., La. & Okla.
Oriental Bittersweet	*C. orbiculata*	May-June	e. Asia; much planted and commonly escaped

outside case orange;
berry within, red

very subject to scale

American Bittersweet

Note:
Flowers and fruit of
American Bittersweet
are terminal
(at ends of stems only);
those of Oriental B.
are axillary
(along the stems).

outside case yellow;
berry within, red

Oriental Bittersweet

MP
186

V

352

American Bittersweet

pical Bittersweet bark   Oriental Bittersweet

MP
187

V

WISTERIA — *Wisteria*   usually Perfect   High-climbing

American Wisteria	*W. frutescens*	May-July	Va. to Fla. & Ala.

(*W. macrostachya*, sometimes considered a variety of American Wisteria, has larger flower clusters and is found in a more westerly range: Ky. to Mo., s. to La. & Tex.)

Chinese Wisteria	*W. sinensis*	May-June	China; much planted and escaped
Japanese Wisteria	*W. floribunda*	May-June	Japan; much planted and escaped

lavender flowers, but varieties and hybrids range from white to rose and purple

← smooth pods

velvety pods →

Note: Flowers of Japanese Wisteria open gradually, those at the tip last. Flowers of Chinese W. all open at approximately the same time; otherwise they are similar.

Japanese Wisteria

American Wisteria

pods typical of both Chinese and Japanese Wisteria

MP 188

V

354

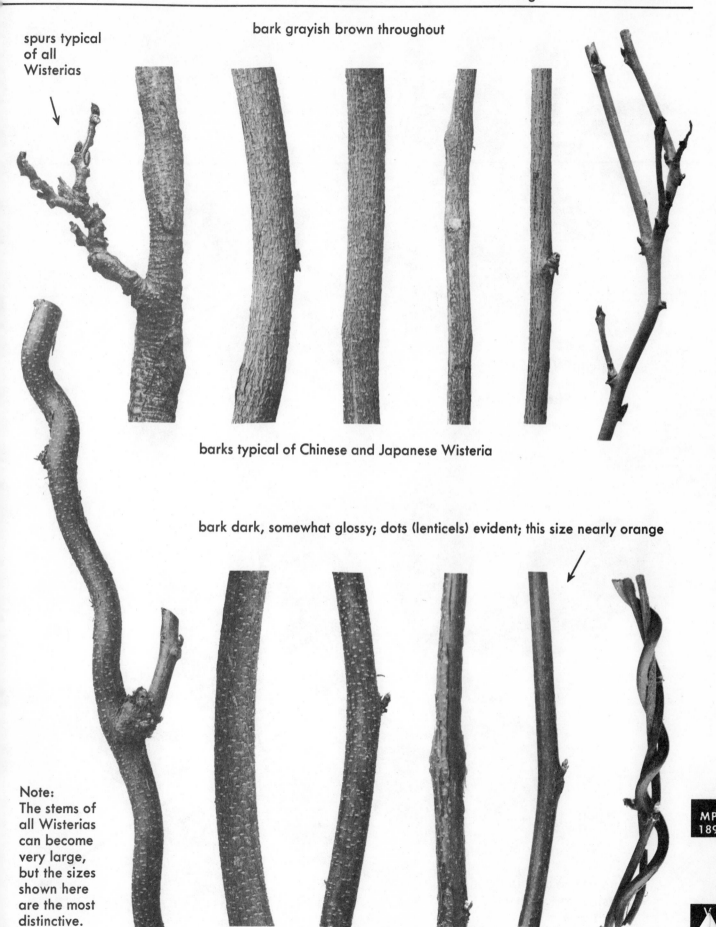

spurs typical of all Wisterias

bark grayish brown throughout

barks typical of Chinese and Japanese Wisteria

bark dark, somewhat glossy; dots (lenticels) evident; this size nearly orange

Note:
The stems of all Wisterias can become very large, but the sizes shown here are the most distinctive.

MP 189

American Wisteria

WISTERIA (con't.)

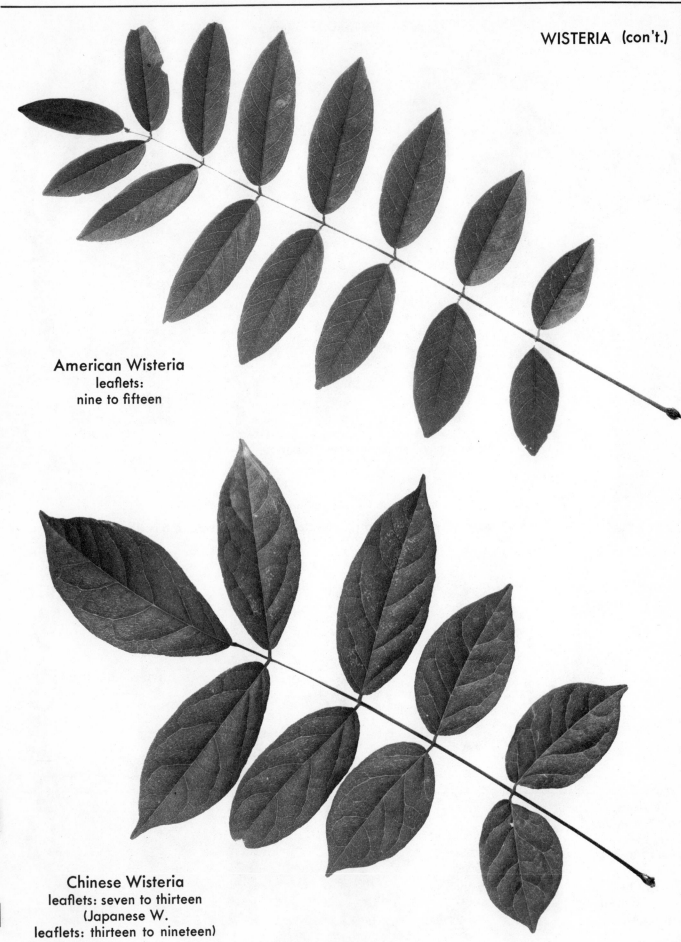

American Wisteria
leaflets:
nine to fifteen

Chinese Wisteria
leaflets: seven to thirteen
(Japanese W.
leaflets: thirteen to nineteen)

MP
190

V

NIGHTSHADE — *Solanum dulcamara*
(Bitter-sweet)

Perfect   June-Aug.   Europe, n. Africa, e. Asia; naturalized in much of N. Amer. Trailing or occ. twining to 8-10 ft.

flowers purple
with yellow centers

It is well worth while
using a magnifying
glass to see the details
of this flower.

berries red
when ripe

The stems and leaves have
a distinctly unpleasant
odor when crushed.

MP
191

V

CURRANT — *Ribes*   (see MP 30)   Perfect

Skunk Currant	*R. glandulosum*	May-Aug.	Lab. to B.C., s. to s. N.E., Pa., W. Va., mts. of N.C., n. Ohio, Mich., Wis. & Minn. Most parts of this plant have a strong skunk odor when crushed. Stems trailing, with ascending branches to 2 ft.
Swamp Red Currant	*R. triste*	April-June	Lab. to Alaska, s. to s. N.E., N.Y., Pa., W. Va., Ill., Ia. & west. Low arching or creeping stems

flowers erect;
white, with
pinkish tinge

flowers drooping;
salmon or dull pink

smooth
red berries

bristly
red berries

red buds

reddish
leaf-stalk

Skunk Currant

Swamp Red Currant

DEWBERRY — *Rubus*   (see MP 46)   Perfect   Prostrate trailing, or scrambling over low objects

Common Dewberry	*R. flagellaris*	May-June	Me. to Man., s. to Va. & Mo.; deciduous
Swamp Dewberry	*R. hispidus*	June-July	N.S. to Mich., s. to Ga. & Ill.; semi-evergreen

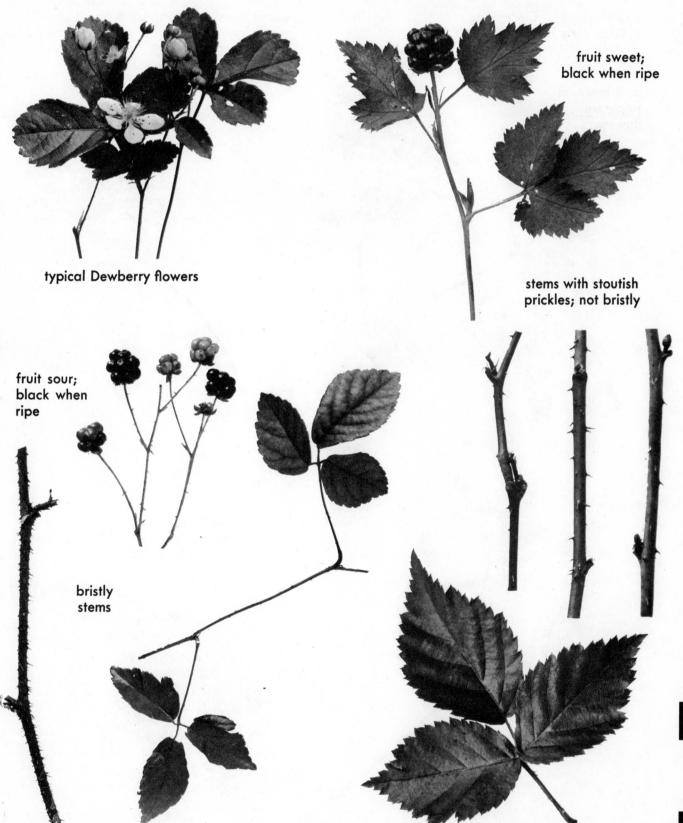

fruit sweet;
black when ripe

typical Dewberry flowers

stems with stoutish
prickles; not bristly

fruit sour;
black when
ripe

bristly
stems

Swamp Dewberry   Common Dewberry

MP
193

V

CLEMATIS — *Clematis*   Perfect (rarely Dioecious)

Purple Clematis	*C. verticillaris*	May-June	Me. & Que. to Man., s. to Md., W. Va., Ohio, Mich. & n.e. la. Climbs to 10 ft.
Sweet Autumn Clematis	*C. paniculata*	Aug.-Oct.	Japan; much planted & escaped here Climbs to 30 ft.
Virgin's Bower	*C. virginiana*	July-Sept.	N.S. to Man., s. to Ga., La. & Kans.
(Traveler's Joy; Love Vine)			Climbs to 20 ft.

pale lavender
to purple
flowers

Purple Clematis

white
flowers

MP
194

V

Virgin's Bower

white
flowers

Sweet Autumn Clematis

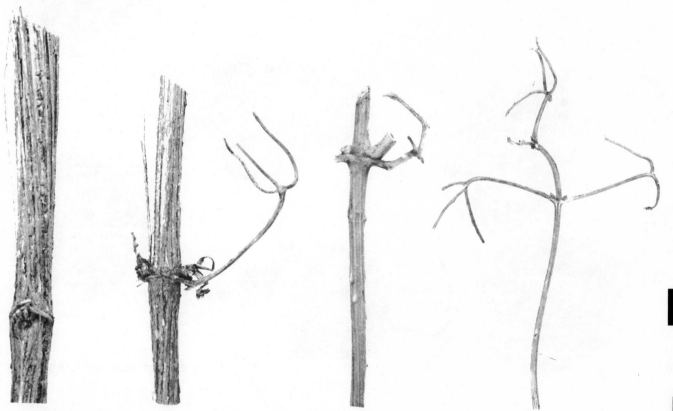

Typical Clematis bark and twigs, showing remnants of leaf-stalks used as tendrils for climbing.

MP
195

V

CLEMATIS (con't.)

Note: Clematis climbs by using its leaf-stalks to grip available supports, instead of having separate tendrils.

Note: Purple Clematis sometimes has no teeth along edges of leaflets.

The leaves of Sweet Autumn Clematis almost never have teeth.

The leaves of Virgin's Bower almost always have teeth.

Purple Clematis

MP 196

V

Sweet Autumn Clematis

Virgin's Bower

EUONYMUS — *Euonymus* (see MP 80)   usually Perfect

Running Strawberry-bush	*E. obovata*	May	w. N.Y., s. Ont. & Mich., s. to Md., Tenn. & Mo. Trailing stems; branches ascending to 12 in.
Wintercreeper	*E. fortunei*	June-July	China; evergreen

More commonly found here in the following varieties:

Common (or Small-leaved) W.   *E. f. radicans*   Climbing, trailing and mat-forming
Bigleaf W.   *E. f. vegeta*   Climbing, trailing or somewhat bushy; heavier than Common W.

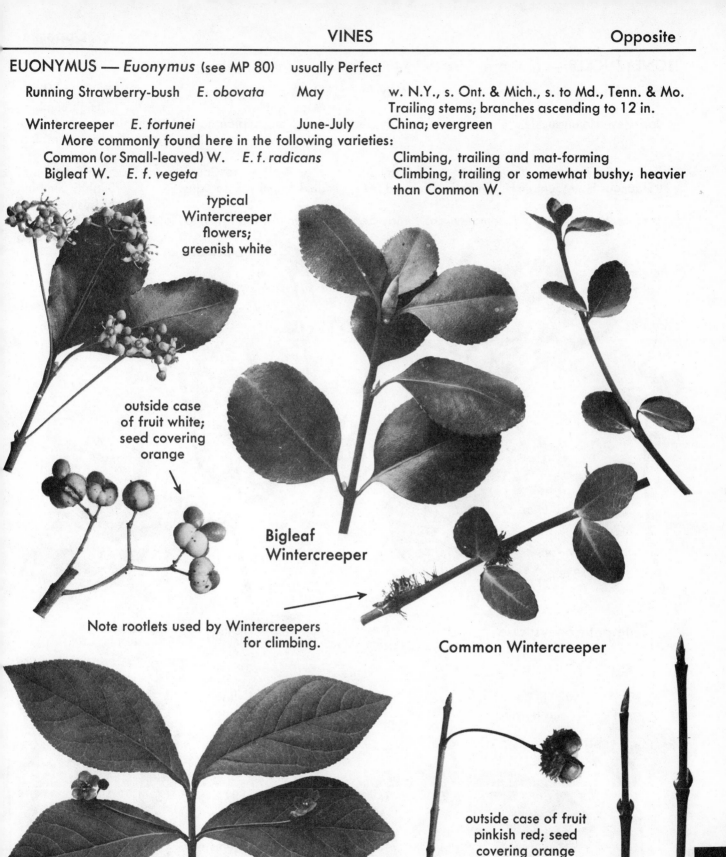

typical
Wintercreeper
flowers;
greenish white

outside case
of fruit white;
seed covering
orange

Bigleaf
Wintercreeper

Note rootlets used by Wintercreepers
for climbing.

Common Wintercreeper

outside case of fruit
pinkish red; seed
covering orange

flowers greenish purple

Running Strawberry-bush

MP
197

V

HONEYSUCKLE — *Lonicera*   (see MP 146)   Perfect   Climbs by twining

Hairy Honeysuckle *L. hirsuta*	May-July	Que. to Sask., s. to w. Vt., N.Y., Pa., Ohio, Mich., Minn. & Neb. High-climbing
Japanese Honeysuckle *L. japonica*	June-Sept.	Asia; much planted & commonly escaped here; semi-evergreen High-climbing, trailing and mat-forming
Limber Honeysuckle *L. dioica* (Glaucous H., Mountain H., Smoothed-leaved H.)	May-June	Me. & Que. to Sask., s. to Ga., e. Tenn., n.e. Ill. & Ia., in var. to Kans. Slightly climbing or bushy
Trumpet Honeysuckle *L. sempervirens*	May-Sept.	Me. to Mich. & Ia., s. to Fla. & Tex.; semi-evergreen High-climbing

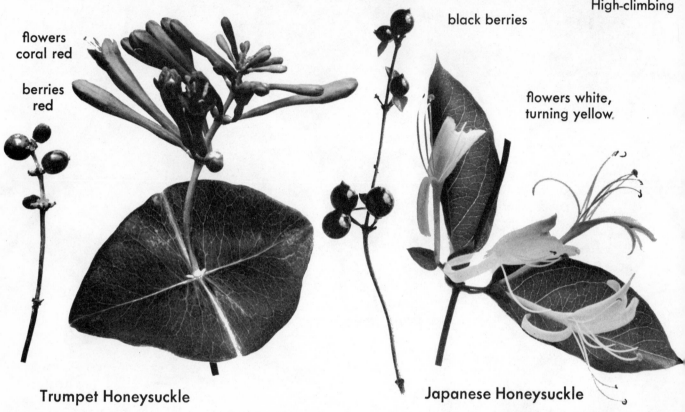

flowers coral red

berries red

black berries

flowers white, turning yellow.

Trumpet Honeysuckle

Japanese Honeysuckle

red berries

flowers pale yellow (sometimes turning orange)

Note: All Honeysuckle vines have hollow stems.

Limber Honeysuckle (Hairy H. similar)

Limber Honeysuckle (Hairy H. similar, but slightly larger and very deep yellow)

barks typical of vine Honeysuckles

MP 198

V

364

uppermost leaves connate
(two leaves joined together)

uppermost leaves
not connate

leaves
whitish
beneath

Note:
There is a
variety with
yellow-veined leaves.

Japanese Honeysuckle

cut-leaf variety
of Japanese H.
occasionally
encountered

Trumpet Honeysuckle
(high-climbing)
(Limber H. similar;
slightly climbing or bushy)

upper leaves connate;
leaves green and hairy on both sides;
hairs especially evident on leaf edges

Hairy Honeysuckle

Hairy
Honeysuckle

Limber
Honeysuckle

MP
199

V

365

TRUMPET-VINE — *Campsis radicans*   Perfect   July-Sept.   Pa. to la., s. to Fla. & Tex.;
(Trumpet-creeper; Trumpet Flower; Cow-itch)   naturalized n. to Conn. & Mich.
High-climbing

flowers orange red

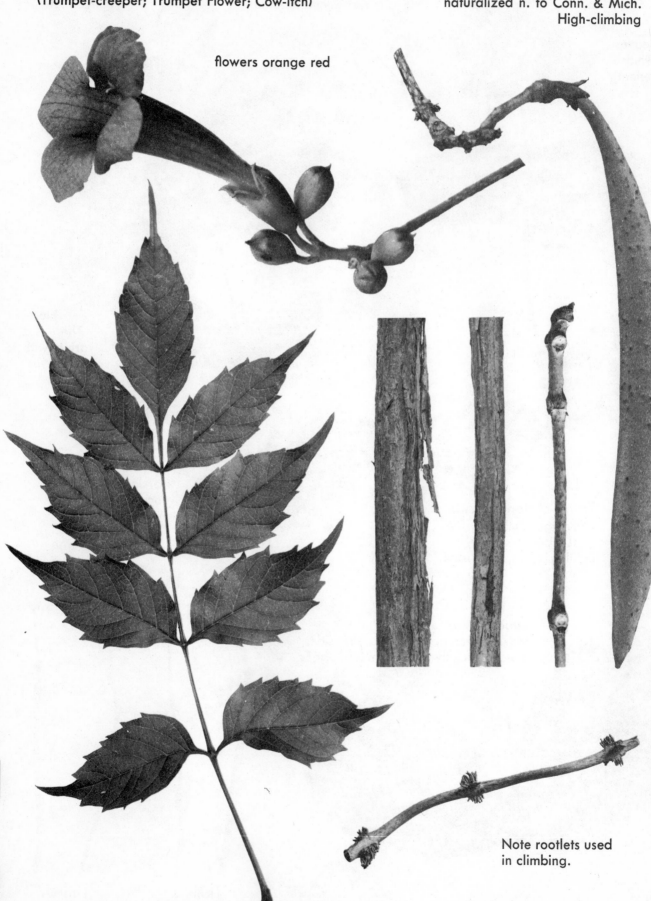

Note rootlets used
in climbing.

MP
200

V

BIBLIOGRAPHY

AMMONS, NELLE. *Shrubs of West Virginia.* (Herbarium, Contribution No. 55.) Morgantown: West Virginia University, 1950.

BILLINGTON, CECIL. *Shrubs of Michigan.* 2nd ed. Bloomfield Hills, Mich.: Cranbrook Institute of Science, 1949.

BLAKE, S. F. *Guide to Popular Floras of the United States and Alaska.* Bibliographical Bulletin No. 23, U.S. Department of Agriculture, 1954.

CORE, EARL L. and AMMONS, NELLE. *Woody Plants in Winter.* Pittsburgh: Boxwood Press, 1958.

DEAM, CHARLES C. *Shrubs of Indiana.* (Indiana Conservation Department Publication No. 44.) Indianapolis: W. B. Burford, 1924.

DOLE, E. J. *The Flora of Vermont.* 3rd rev. ed. Burlington, Vt.: Free Printing Press, 1937.

FASSETT, NORMAN C. *Spring Flora of Wisconsin.* 3rd ed. Madison: University of Wisconsin Press, 1957.

FERNALD, M. L. *Gray's Manual of Botany.* 8th ed. New York: American Book Co., 1950.

FREEMAN, OLIVER M. *Annotated List of the Plants Growing Naturally at the National Arboretum.* (National Arboretum Contribution No. 1.) Washington, D.C.: Government Printing Office, 1953.

GRAVES, ARTHUR HARMOUNT. *Illustrated Guide to Trees and Shrubs.* Wallingford, Conn.: published by the author, 1952.

GRIMM, WILLIAM CAREY. *The Shrubs of Pennsylvania.* Harrisburg, Pa.: Stackpole Co., 1952.

HARRINGTON, HAROLD D. *Keys to the Woody Plants of Iowa.* University of Iowa Studies in Natural History, Vol. 17, No. 9. Iowa City: The University, 1940.

HERMANN, FREDERICK J. *A Checklist of Plants in the Washington-Baltimore Area.* 2nd ed. Washington, D.C.: Smithsonian Institution, 1946.

HODGDON, ALBION R. and STEELE, FREDERIC L. *The Woody Plants of New Hampshire.* (Bulletin 447, Agricultural Experiment Station.) Durham, N.H.: University of New Hampshire, 1958.

HOUSE, HOMER D. *Annotated List of the Ferns and Flowering Plants of New York State.* Albany: University of the State of New York, 1924.

HYLAND, FAY and STEINMETZ, FERDINAND H. *The Woody Plants of Maine.* Orono: University of Maine Press, 1944.

MASSEY, ARTHUR B. *Evergreen Shrubs, Vines and Ground Cover Native in Virginia.* Blacksburg: Virginia Polytechnic Institute, 1952.

REHDER, ALFRED. *Manual of Cultivated Trees and Shrubs.* 2nd ed. New York: Macmillan Co., 1940.

ROSENDAHL, CARL OTTO. *Trees and Shrubs of the Upper Midwest.* Minneapolis: University of Minnesota Press, 1955.

SHANKS, ROYAL E. *Checklist of Woody Plants of Tennessee.* Knoxville: University of Tennessee, 1952.

STONE, WITMER. *The Plants of Southern New Jersey.* Trenton: 1911.

TEHON, LEO R. *Fieldbook of Native Illinois Shrubs.* Urbana: Natural History Survey Division, State of Illinois, 1942.

GLOSSARY

Appressed—close to (not sticking out); opposite of divergent.

Axillary—(as buds) borne in angle or axil formed by two diverging parts.

Bloom—whitish, usually powdery, covering that can be rubbed off.

Bract—a modified leaf, usually at the base of flowers or leaves.

Calyx—outer part of a flower (at the bottom when the flower is open). It is usually green (but occasionally colored, especially when a flower produces no colored petals).

Compound—having two or more similar parts; a leaf made up of several leaflets on the same leaf-stalk is a compound leaf.

Deciduous—not permanent, falling off, as with leaves in autumn.

Dioecious—male flower on one plant, female on another.

Divergent—sticking out; opposite of appressed.

Genus—main name of a plant (Dogwood, Honeysuckle, etc.).

Hybrid—a cross between closely related plants.

Leaf-scar—a scar marking the point of separation from the twig when a leaf drops off in the fall.

Leaflets—the single blades of a compound leaf.

Lenticels—openings on the surface of leaves or bark allowing the passage of air or gases; these appear as dots on the surface.

Lobe—the part of a leaf sticking out when the margins are not uniform. This is the opposite of the sinus, which is the cut-out part of an uneven leaf.

Monoecious—male and female flowers distinct from one another, but on the same plant.

Mucro—short, abrupt tip.

Node—the point on the twig that bears a leaf or leaves.

Palmate—spreading from one point; a palmately compound leaf is one where all the leaflets branch from one point.

Perfect—both male and female parts in the same flower.

Petiole—leaf-stalk.

Pinnate—having parts arranged along two sides; a pinnately compound leaf has leaflets along both sides of the leaf-stalk.

Pistil—the seed-bearing (female) part of a flower.

Pistillate flower—female flower.

Pith—inner part of a twig or stem that is not woody but pulpy.

Polygamous—bearing both perfect and unisexual flowers on the same plant.

Raceme—elongated flower (or fruit) cluster, the individual flowers (or fruit) arranged along a single axis or stem.

Rachis—the continuation of the leaf-stalk of a compound leaf, beginning at the lowest leaflet(s) and continuing to the end leaflet.

Revolute—rolled backward from the margins or the apex.

Sepals—divisions of the calyx (sometimes petal-like).

Sessile—without a stalk.

Simple—applied to a leaf with only one blade.

Sinus—the cut-out part of a leaf with uneven margins; the opposite of lobe.

Species—a particular kind of plant within a given genus, such as Flowering Dogwood (Dogwood being the genus).

Stamen—the pollen-bearing (male) part of a flower.

Staminate flower—male flower.

Sterile—incapable of bearing fruit.

Stipules—leaf-like objects at the base of leaf-stalks.

Sub—prefix indicating "somewhat," "slightly" or "partly."

Subgenus—a subdivision of a genus.

Subshrub—a term used to indicate a plant intermediate between a perennial and a true woody shrub. The base is always woody.

Whorl—a circle of three or more similar parts around a central point, as three or more leaves growing around a twig from one joint, or node.

INDEX
TO
MASTER PAGES

INDEX TO MASTER PAGES

Matrimony-vine, 125
  Chinese
  Common

Mayflower (Trailing Arbutus),
  169B

Meadowsweet (Spirea), 40
  Birch-leaved
  Broad-leaved

Mealberry (Common Bearberry),
  164A

Menispermum canadense, 184

Menziesia, 103
  Allegheny

Menziesia pilosa, 103

Mezereum (Daphne), 91B

Minniebush (Menziesia), 103

Mistletoe, American, 17A

Mitchella repens, 168A

Mock-orange, 22
  Large-flowered
  Scentless

Moonseed, Common, 184

Moosewood (Leatherwood),
  91A

Moosewood (Viburnum), 136
  (Hobble-bush)

Moss, Flowering, 165C

Mountain Ash Spirea (False
  Spirea), 24

Mountain Heath, 161A

Mountain Lover (Pachistima),
  164B

Myrica, 4
  M. cerifera
  M. gale
  M. heterophylla
  M. pensylvanica

Myrtle (Periwinkle), 171B

Myrtle, Sand-, 113B

Myrtle, Sea (Groundsel-bush),
  155

Myrtle, Wax-, 4

Nanny-berry (Viburnum), 136

Nemopanthus mucronatus, 83

New Jersey Tea, 45A

Nightshade, 191

Ninebark, 27

Oak, 14
  Chinquapin
  Scrub

Oil-nut (Buffalo-nut), 16

Oleaster, 96

Olive, Russian (Oleaster), 96

Opossum-wood (Silverbell), 128

Pachistima, 164B

Pachistima canbyi, 164B

Pachysandra, 171A

Pachysandra, 171A
  P. procumbens
  P. terminalis

Parthenocissus, 178
  P. quinquefolia
  P. tricuspidata

Partridge-berry, 168A

Pepperbush, Sweet (Clethra),
  102

Pepper-Vine (Ampelopsis), 173

Periwinkle, 171B

Philadelphus, 22
  P. coronarius
  P. grandiflorus
  P. inodorus

Phoradendron flavescens, 17A

Phyllodoce coerulea, 161A

Physocarpus opulifolius, 27

Pieris, 114
  P. floribunda
  P. japonica

Pigeonberry (Sarsaparilla), 92

Pimbina (Viburnum), 136
  (Highbush Cranberry)

Pinxterbloom (Azalea), 104

Pipsissewa, 170B
  Striped

Plum, 54
  Beach
  Red
  Wild
  Wild Yellow

Poison Ivy (Sumac), 74

Polygonum cuspidatum, 17B

Porcelainberry (Ampelopsis),
  173

Possum-haw (Viburnum), 136
  (Smooth Withe-rod)

Potentilla, 64
  P. fruticosa
  P. tridentata

Prickly Ash, 70

Prince's Pine (Pipsissewa), 170B

Privet, 132
  California
  Common
  Ibota
  Regal

Prunus, 54
  P. americana
  P. maritima
  P. pumila
  P. virginiana

Ptelea trifoliata, 71

Pyracantha coccinea (lalandii),
  42

Pyrularia pubera, 16

Pyxidanthera barbulata, 165C

Pyxie (Flowering Moss), 165C

Quercus, 14
  Q. ilicifolia
  Q. prinoides

Raspberry, 46
  American Red
  Black
  European Red
  Purple Flowering

Rat Stripper (Pachistima), 164B

Ratbane, Spotted (Pipsissewa),
  170B

Redbud, 68

Rhamnus, 84
  R. alnifolia
  R. cathartica
  R. frangula
  R. lanceolata

Rhododendron (104); 108
  Carolina
  Catawba
  Lapland
  Mountain
  Purple
  Rosebay